The Virtual Engineer™

The Virtual Engineer™
21st century product development

By Howard C. Crabb, Ph.D.

Society of
Manufacturing Engineers
Dearborn, Michigan

American Society of
Mechanical Engineers
New York, New York

Copyright © 1998 by Howard C. Crabb and Society of Manufacturing Engineers

987654321

All rights reserved, including those of translation. This book, or parts thereof, may not be reproduced in any form or by any means, including photocopying, recording, or microfilming, or an information storage and retrieval system, without written permission from the copyright owners.

No liability is assumed by the publishers with respect to the use of information contained herein. While every precaution has been taken in the preparation of this book, the publisher assumes no responsibility for errors or omissions. Publication of any data in this book does not constitute a recommendation or endorsement of any patent, proprietary right, or product that may be involved.

No responsibility is assumed by the publisher for any injury and/or damage to persons or property as a matter of product liability, due to negligence or otherwise, or from any use or operation of any methods, products, instructions, or ideas contained in the material herein.

Library of Congress Catalog Card Number: 97-061866
ISBN: 0-7918-0066-0

Additional copies may be obtained by contacting:

American Society of Mechanical Engineers
22 Law Drive, P.O. Box 2900
Fairfield, NJ 07007-2900
1-800-843-2763

SME staff who participated
in producing this book:

Donald A. Peterson, Senior Editor
Rosemary Csizmadia, Production Team Leader
Jennifer Courter, Production Assistant
Sandra Cavazos, Production Assistant
Christine Verdone, Cover Designer

Printed in the United States of America

800660

To JoAnn, my wife, for her love, patience, support, and encouragement during my long career, and to my children, Paul, Kathleen, and Thomas for their enterprise, originality, and eternal optimism.

Contents

Industry Forecast: Dramatic Change xi

Orders-of-magnitude Process
  Improvement xiii

Foreword xv

Preface xix

Acknowledgments xxiii

Chapter 1.
An Environment of Change 1
  The Way We Were 1
  The Dynamics of Change 2
  The Impact of Information Technology 6
  A Systems Approach to Productivity 8

Chapter 2.
The Elements of Change 13
  Time as a Change Agent 14
  Flexibility to Accommodate Change 15
  Categorizing Benefits 17
  Realizing Benefits 28
  So What! 35

Chapter 3.
Technology as a Tool 37
  Microprocessor Technology 37
  Semiconductor Investment Drivers 51
  Engineering Workstation Price and Performance 52

Software 53
So What! 54

Chapter 4.
Reinventing the Current Process 57

Information as an Economic Asset 57
Enter Visual Engineering 59
Who's Listening to the Customer? 63
Who Understands the Process? 65
The Product Development Process 65
The Way of Success 86

Chapter 5.
2000 and Beyond 89

Product Development Practices of Yesterday 89
The Enterprise as a System Runs on Information
   Technology 91
Next-generation Computer-aided Technologies—
   They're Visual 91
Challenges and Opportunities 99
Next Generation Product Development 104
Systems Engineering—Integrator of Process, People,
   and Technology 107
What to Look For 109

Chapter 6.
Organization Overhaul 111

Integrate, Innovate, or Evaporate 112
The Past is Prologue 113
Enterprise Thinking 115
Technology, Organizations, and People 116

Chapter 7.
Best Practices 125

Pathway to Prosperity 125
Tomorrow and Beyond 162
The Next Paradigm Shift—Where to Look for
   Emerging Changes 163

Index 165

Industry Forecast: Dramatic Change

Computer-Aided Engineering (CAE) refers to the use of computers in the engineering process. CAE systems are used by engineers throughout the new product development process to design, depict, analyze, manufacture, and inspect components, assemblies, tools, and fixtures.

In recent years, there has been a rapid expansion in CAE, driven by several factors:

- Most problems are system related, not people related.

- More powerful computers (especially workstations and desktop PCs).

- More capable software (especially for solids modeling and finite element analysis).

- New manufacturing technologies (especially for rapid prototyping).

These technological developments have the collective potential to bring about fundamental changes in the way engineering is done.

During the same time frame, the business environment for the engineering/manufacturing enterprise has undergone dramatic and rapid shifts: globalization of markets, heightened competitive pressure, increasing product complexity, and decreasing product cycles.

The result is a compelling requirement for change in today's manufacturing enterprise. It must reinvent itself to bring about major changes in its business processes in general, and in its new product development processes in particular. In so doing, it must quickly and effectively understand, acquire, and deploy CAE technology, which is itself evolving.

Dr. Crabb's book is a welcome and useful aid in facing this daunting challenge. Drawing on his considerable experience, he explains, in clear and effective prose:

- How change benefits the organization, in terms of reduced cycle time, reduced engineering changes, reduced number of physical prototypes, and improved product quality.

- Developments in the semiconductor industry and their impact on the desktop computing environment.

- The steps required to reinvent the current new product development process.

- The best practices of existing firms.

This book will be a valuable guide for the practitioner and an important addition to the literature on Computer-Aided Engineering.

David C. Gossard
Professor of Mechanical Engineering
Massachusetts Institute of Technology

Orders-of-magnitude Process Improvement

Howard Crabb's vision suggests a new type of engineer who will flourish in tomorrow's environment. Therefore, this book will be of great interest to engineering educators and students. Likewise, research engineers in academia, industry, and government will find that the procedures Howard advocates will improve the process of research in the same way they have improved the product development process. I can speak from personal experience in affirming that electronic prototyping technologies such as visual engineering, solid modeling, knowledge-based engineering, object modeling, multiphysics simulation, and parallel processing can lead to orders-of-magnitude improvement in research productivity and relevance.

Dr. Crabb established a reputation as a corporate visionary during a long and distinguished career in engineering management in the automobile industry. His ability to foresee trends in technology and discern their eventual impact upon product development led to many revolutionary innovations in the engineering design process. In this book, he describes these ideas and presents his view of the future. Engineering managers will be rewarded with an essential road map for understanding and dealing with a rapidly changing technical and competitive environment. As Howard remarks, manufacturing organizations are in a race between "the quick and the dead."

Thomas J. R. Hughes
Professor of Mechanical Engineering
Chairman, Division of Mechanics and Computation
Stanford University

Foreword

This book describes the quest for changing the product development process and the new breed of engineer that is evolving to develop improved or new products. It describes the framework for change management of today's product development process and steps for getting started.

The industrial world is undergoing fundamental changes in the way business is done as the 21st century emerges. The Industrial Age is being replaced by the Information Age and drivers for successful businesses are radically different from those of past decades. The traditional systems approach of past decades used computer-literate professionals to examine work processes and separately automate them with computer technology—both hardware and software—creating "islands of automation." The emphasis for success now is the process—a process that results in quality products at a reasonable cost and in a short period of time.

Once, modeling and analysis occurred late in the product development process, usually after the design was documented on a CAD system, and then, only analysts, not design engineers, were empowered to use the tools. Today this would lead to extinction.

It was in the early '80s that Ford began changing the process—first moving solids modeling and analyses, such as finite element analysis and thermal kinematics, to the beginning, and then empowering design engineers with these new tools. This became known as predictive engineering. It began with the design of mechanical components and assemblies, and grew to encompass full vehicle design.

Prediction was next expanded to defining manufacturing processes in a collaborative R&D project called Rapid Response Manufacturing (RRM) under the management of the National Center for Manufacturing Sciences, sponsored by NIST, and involving a number of automotive, aerospace, electronics, and software companies.

Many leaders and innovators worked long and hard to ensure Ford's leadership in this strategically important technology. Among them were Bill Powers, Vice President R&D; Gordon Willis, Chief Engineer, Powertrain; Bob DeLosh, Systems Office Manager, Powertrain Operations; Peter Sferro; Wayne Hamann; Dick Radtke; and Raj Birla.

Howard Crabb, formerly Manager of Computer Aided Engineer Functions at Ford, Alpha Group, was instrumental in implementing these ideas, which continue to make a major impact on Ford. With a virtual team recruited from various line organizations, Howard's development projects resulted in good technology and an ever-expanding population of "virtual engineers." His insight into the implications of the latest hardware and software technologies, and a lifetime of experience in information systems, have made his message one that senior management should listen to carefully.

Real process improvement comes from using systems engineering to optimize the whole corporate enterprise for quality products at reduced cost and faster time-to-market. Systems engineering is a return to basics that stresses understanding all segments and their interrelationships. In other words, predictive engineering has evolved from the part, to the assembly, to the product, to the manufacturing process, and now to the enterprise.

Information technology, in the form of visual and predictive engineering, is drastically changing the way products are designed and built. Visual engineering is a key enabler of simultaneous engineering and allows everyone in an organization to see and understand information in the same way. Incorporating predictive engineering permits concepts to be evaluated in minutes or hours, rather than days or months. The process integrates product development, manufacturing, marketing, and costing into a single new engineer—the virtual engineer.

The Virtual Engineer™ describes the successful processes and types of individuals that will follow. Product development, the kernel of manufacturing companies, must undergo fundamental changes and return to the basics to be best in class in the next decade. It describes reinventing a new product development process for engineered products which uses information technology to continuously enhance the entire enterprise. With the rapid changes occurring today in computer hardware, continuous improvement must be a vital link that corporations must include to meet their short- as well as long-term strategies.

In the very near future, the ability to architect and deploy such complex systems will distinguish the winners from the losers.

Larry McArthur
President and CEO
Ascent Logic Corporation

Preface

This book discusses, in business terms, reinventing the new product development process for engineered products. Though many of the examples given in the book are taken from the automotive industry, they are applicable to any company involved in the manufacture of an engineered product.

The new product development process has long been the private domain of engineers. It has been only recently that management has begun to ask the tough questions as to how this process supports time-to-market, and most importantly, *time-to-money* for new products. In this text I describe major elements of product development, including:

- Issues facing all companies and benefits of change,
- Evolution of computer-based technology used in the process,
- Discussion of actions for reinventing the process,
- An overview of best practices, with case study examples,
- Trends in product development,
- Call for action and steps for getting started.

Pervading this text is a theme of returning to the basics of systematically applying organizational structure, people, processes, and technology to the improvement of engineered products. Most companies have good people, excellent product and process technologies, and certainly an external push from markets and competition to get better. The key for these companies is to step back and take a strategic systems engineering approach to integrating the people, product, process, and technologies in the business of new product development.

To reinvent product development processes, management must abandon its current notions about organization and work flow. In the following pages I describe a new business model for developing new products. This model also has an associated set of techniques that executives and managers will need to use to reinvent a repeatable process that allows the highest quality products to be developed in the shortest period of time and at a reduced cost. Some of the management themes of this new business model are:

- Global business competition dictates that product development must be a core competency for manufacturing companies and their suppliers.

- At least 50 percent of the intellectual property and product development experience will be leaving major companies over the next decade.

- Process and product information management must take precedence over technology in the quest for improved product development.

- High-performance computing is becoming very affordable and is available today for deployment on the engineer's desktop.

- The evolution from physical prototypes to electronic and mathematical prototypes is the driver for reducing product development process time.

- The fully-burdened life-cycle cost of a product development engineer at many companies is more than $12 million in a typical 30-year career.

- The bottom line is that most companies must double or even triple their product development and engineering productivity by the year 2010.

The manufacturing companies enjoying the greatest success are those that have evolved from an engineering-driven product development process to a customer-driven process. Today, most new product engineering enhancements have their roots in the "voice of the customer," as captured from interviews, surveys, focus groups, and numerous trade shows. Business and technology issues facing manufacturing companies today include:

- Need to compress product development times from a high of 48 months to less than 24 months.

- Evolution from time-consuming and expensive physical prototypes to rapidly-developed and low-cost electronic prototypes.

- Evolution from outdated two-dimensional CAD/CAM design practices to a three-dimensional, CAE-driven solids modeling design process.

- Increased focus on net profits per production unit as opposed to increased market share in an industry that is rapidly maturing.

Information technology and *visual engineering* will drastically change the way we design and build products in the future. Visual engineering is the representation of design concepts and predictive engineering simulation as realistic functioning objects within their systems environment and predictive engineering is the mathematical simulation of time-based occurrences. Visual engineering allows all with a need to know to see and understand information in the same way. The information can be in the form of a solids model of a new product design or geographic displays of data searches and analysis on the buying habits and other characteristics of customers in a particular region.

Desktop computer performance is doubling every 12 months. Tasks we find difficult to do with computers today will become routine in the future. Moreover, the number of home-based computers has finally outnumbered office-based computers, creating a new generation of people experienced in the use of these tools for performing a wide variety of tasks. The successful organization in the future must be multidisciplined and focus on developing innovative product and process improvements throughout newly-structured organizations based on flexibility and fluidity.

Some of the improvements expected from this philosophy change include:

- A virtual collapse of the communication walls separating engineering, manufacturing, support organizations, and supplier networks.

- Embedding of robust manufacturing and design practices into a direct engineering environment driven by CAE and visual engineering.

- Increased use of web-based electronic bookshelves that contain proven design objects that can be reused for building new product designs.

- Productivity increases of 10, 100, and even 1000 times greater than the 10 to 50 percent improvements today.

This book is intended to help senior management in its quest to reinvent the product development process. It helps describe the framework for change management of today's processes and steps for getting started.

Howard C. Crabb, Ph.D.
Interactive Computer Engineering
1997

Acknowledgments

This book emerged from my experiences with product development processes, and how new technologies could change the way engineering simulation was performed as well as impact industrial management structures.

I would like to thank two people who helped initiate the endeavors that ultimately led to this book. Without the vision and support of W. F. Powers and F. G. Willis, many of the process changes I describe would still be in discussion stages.

The ability to evaluate changes to engineering product and process development was under the charter of the Alpha Group within Ford. I was privileged to be a member of this organization and would like to thank J. A. Manoogian and D. F. Hagen for their many constructive suggestions. Another very special contributor was W. W. Boddie, who was not only a champion, but who always found a way for the processes to be evaluated when the standard reply was "you can't do that." The product and manufacturing knowledge he imparted to me will always be appreciated.

Many important contributors were my friends and colleagues who worked with me to develop and implement some dramatically new ideas. Thanks to W. C. Hamann for his support. Special thanks to R. R. Radtke for his continuous support and ability to foresee next-generation software, workstation, and supercomputer evolution; his continuous support, enthusiasm, and focus is greatly appreciated. Thanks also to R.T. Wong for implementing information technology changes on over 10,000 desktops. He was truly a valued member of our team. There were many others who played important roles and to whom I am indebted: those in manufacturing, product engineering, product planning, purchasing, and marketing, as well as many other organizations.

Special thanks go to the suppliers of hardware, software, and the providers of engineering services who embraced the emerging new processes and supported the development of unique engineering talent to use new mathematical simulation software. Computer hardware was changing rapidly, and contributors I would like to thank include R. J. Paluk, S. Wallach, and W. P. Roelandts. Companies helping to define new visions for hardware included Hewlett-Packard Co., Convex Computer Corp., Silicon Graphics, Inc., Cray Research, Inc., IBM, Apollo Computer Co., and Alliant Computer Co.

The evolving engineering process required a new type of engineer who could do traditional predictive engineering, but also begin to simultaneously integrate manufacturing processes. Thanks to P. Krishnaswamy of EASi Engineering, Inc., and G. W. Fisher of Fisher Unitech, Inc. for their participation in the effort to obtain and train that talent.

Software, the heart of systems, was the key enabler and included revolutionary changes called visual engineering and solids modeling. Wavefront Technologies, Alias Corporation, and Evans and Sutherland, Inc. were pioneers in visual technology and helped immensely. Special thanks to W. Davison, B. Enright, and R. Pawlicki who helped mature the technology in the engineering environment. Thank you also to J. S. Smith and K. Borchert for communicating our message with outstanding video presentations that were shown globally.

Solids modeling was considered revolutionary, and Aries Technology, especially Larry McArthur, played a pivotal role in introducing this technology, commercializing it as a worldwide industry standard. The ability to design from cradle to grave was possible due to his constant commitment to open standards for geometric data.

As one who started out in the field, engineering software is close to my heart. I would like to thank ANSYS, Inc., Centric Engineering Systems, Inc., Computational Mechanics Co., Inc., Hibbitt, Karlsson and Sorensen, Inc., MARK Analysis Research Corporation, MacNeal-Schwendler Corp., and RASNA, whose important contributions were the development of new predictive engineering techniques, as well as their standardization through commercialization.

Of critical importance to this book were those who worked long hours to clarify early drafts. These include J. F. Gloudeman, R.O.S.E. Informatek Gmbh; D. C. Gossard, Massachusetts Institute of Technology; T. J. R. Hughes, Stanford University; L. McArthur, Ascent Logic Corp.; O. B. Marx III, TMW Enterprises, Inc.; and J. T. Oden, University of Texas at Austin. I am grateful for their help and their insight. Words cannot express my gratitude to R. Fox at Fox and Associates who continually helped clarify many drafts throughout its development.

And to the many others who assisted me in numerous ways to complete this book—my deep appreciation.

Howard C. Crabb, Ph.D.
Interactive Computer Engineering
1997

An Environment of Change

THE WAY WE WERE

Reams have been written over the past decade about the paradigm shifts that must occur for American industry to be competitive in emerging global markets. Being a nation of producers of both capital and intellectual goods, the U.S. has a vital stake in the success of its manufacturing sector.

Starting in the 1950s, manufacturing production elevated the standard of living of Americans to unprecedented levels. The pent-up consumer demand for capital goods after World War II spurred record growth in new industries and created new, well-paying jobs.

In the 1970s, however, a slow but gradual erosion of the nation's standard of living began. Inflation soared as companies struggled to hold prices. Offshore, manufacturing capability soared, and the period marked the opening for foreign competition to come ashore.

The 1990s witnessed an America seemingly content with an economy growing at an annual rate of about 3 percent.

At the same time, Asian countries such as China are just realizing their potential and growing at an average of 15 percent annually.

New tigers such as Taiwan, Korea, and Malaysia are growing at 10 percent a year. Latecomer upstarts such as Vietnam and some Eastern European countries are coming on strong. The pace of change that started in the 1990s is catapulting competing nations into a high stakes race between the quick and the dead.

THE DYNAMICS OF CHANGE

The changes occurring today are predictable, and really began in the 1960s. They are fundamental and affect every aspect of our daily living. Just as the Industrial Age totally transformed the economic and social structure of the world, the Information Age is changing the fiber of economic and social structure today. In the 21st century, the measure of success will be entirely different from what it is today. Past successes of many companies shaped their current organizations and approaches to managing business. Today, new elements must be factored into the equation.

> *Companies must compete on their ability to develop and deliver intellect as opposed to merely the conversion of raw material.*

Recently, a major U.S. automotive company announced quarterly earnings of about $1.4 billion, which seemed enormous. Further analysis, however, revealed that this was less than 5 percent of sales ($35 billion for the quarter). The chairman of the company has now focused on the new product development process to get the company (and its owners) to meet its target of 5 percent return on sales. Many companies are changing from a market share growth focus to net profit growth as a key business metric.

It is time to retire outdated product development concepts and focus on process leadership that is time-driven, knowledge-based, and technology-focused.

The emphasis of this change relates to people and how they develop a product that can be manufactured with the highest quality at the least cost. The automotive industry, as an example, has evolved from an engineering-driven, product-push development process to a customer-driven, market-pull process. Most new enhancements on vehicles today have their roots in the *voice of the customer*, as captured from interviews, competitive benchmarking, surveys, focus groups, and numerous trade shows.

The manufacturing industry has been forced to change. Some companies have embraced change faster than others. The 1980s was a difficult decade for most automotive companies, because of the emergence of the Japanese as a market presence and a formidable competitor.

For the rest of industry in the U.S., the 1980s were equally fraught with change. More than one-half of all companies were restructured. Over 700,000 companies sought bankruptcy protection to keep operating, while more than 450,000 eventually failed.

> *In looking back, if the 1980s are viewed as the decade of change, then the 1990s will probably go down as the decade of instability.*

Change is fact and it is not going away. The infusion of increasingly advanced computer-based technologies will only accelerate its pace. If change is not embraced, and if fundamental change management is not made part of a company's core competency, that company, as we know it today, will not survive.

Competitive Factors

For today's competitive environment, most change is driven by external factors. Michael Porter's landmark book, *Competitive Strategy* (The Free Press, 1985), introduced several external competitive factors that affect most companies involved in the manufacturing of engineered products.

Assessment. The first is that of current competitors. Everything that can be done should be done to assess all of their capabilities as part of your business plan. Competitive assessment has become a clandestine industry in and of itself, and sophisticated means and methods are used to stay abreast of the competition.

Recently, a major defense contractor purchased an unused Soviet-made air defense system for about $50 million, 20 times less than its cost of over $1 billion. This system will be assessed piece by piece to establish its capability. Once this is established, the U.S. Department of Defense will redefine its system requirements for next-generation missiles and manned aircraft. This competitive assessment will result in new weapons that exceed the former Soviet Union's ability to track U.S.-made missiles and aircraft. This is a textbook example showing that competitive assessment is not just limited to taste tests of colas or beers, it is a strategic process.

> *Competitive assessment has been raised to a science as part of an overall change management process.*

Competitive assessment is a change management process that reaches to the core of the organization. Often, the real competition comes from within. The changes explored herein are fundamental and flow to the very base of business thinking and philosophy. Change is needed from the ground up, first in mindset, then throughout the organizational, technological, and information transfer processes.

In many companies, middle management is the key to change. In this tier are the guardians of the gates leading to either the status quo or to behavioral changes that would allow the acceptance of new ideas, thoughts, processes, and techology enablers. Yet middle management is often excluded from the strategy development that results in the call for change. Without ownership of change by this group, most companies are doomed to failure in their change management process.

> *Middle management is the gatekeeper to change in most industrialized companies today. Without their ownership, the change process will fail.*

A major U.S. manufacturer began its effort to change its way of doing business by the year 2000. Top management and high-priced external consultants had crafted a strategy to position the company to world-class status in quality, cost, market share, and employee pride.

Unfortunately, this company did not adequately involve middle management in the strategy and change management deployment. Several years later, the company is still striving to cope with the change management process, and in fact, has lost some market share and key business focus.

Make or buy? The second form of competition is that from customers, in the sense that you may be providing a product or service to them that they might consider producing or doing themselves if they could improve on price, quality, or delivery. This flip side of outsourcing is a real threat to a company unable to meet or better the benchmarks of the industry.

Supplier squeeze. Still another form of competition is that from current suppliers. They could be providing key parts of your product that control your output and could move up the manufacturing "food chain" and cut you completely out of the picture. It is imperative to success to stay constantly alert for technology advances and the efficiencies they can afford in their corporate culture. Continuous improvement must be a core competency.

Technology use. Technology, in the form of both materials and processes, also poses a threat which must be assessed. Are you using it effectively? Is your equipment and the way you are using it cost-effective? Do you have islands of efficiency in a sea of neglected opportunity? Have you assessed your organization in terms of a single system?

Regulations. The regulatory environment, in the form of government rules and laws, is also a major competitive threat that has kept most U.S.-made products out of Japan.

THE IMPACT OF INFORMATION TECHNOLOGY

Henry Ford changed the industrial landscape with automation and specialization of the manufacturing process. This breakthrough in technology enabled Ford to double the wages of his workers, who in turn became customers for the products they were manufacturing and assembling in reduced times and with record efficiencies.

> *Automation of the manufacturing process was the great breakthrough for American companies during the first 60 years of this century.*

Just as Ford changed the industrial landscape for the first 60 years of the 20th century, information technology (IT) will drastically change the way we design and build products in the next century.

In the next millennium, advanced information technology will affect all aspects of an enterprise. Rather than functioning as distinct disciplines, elements of the manufacturing organizations will evolve into an integrated system using seamless information processing and exchange capabilities, and computer-based technologies operating in real time.

At the core of this transformation is *visual engineering*, the representation of design concepts and predictive engineering simulation as realistic objects and time-based occurrences. The information could be a 3-D representation of a new product design in its systems environment or geographic displays of data mining (searches and analysis) on customer buying habits in a particular region. Or, it could be a 3-D model of a planned facility layout with animated movement of materials through the various stages of the manufacturing and assembly process. What makes it possible are gigantic advances in computing power and software flexibility.

> *Visual engineering allows everyone to see and understand information in the same way.*

Computer performance on the desktop is doubling every 12 months. Tasks difficult to do with computer software today will be as easy as using spreadsheet programs by the year 2000 and beyond. In 1997, the number of home-based computers finally outnumbered office-based computers. A new generation of people experienced in the use of these tools routinely perform a wide variety of tasks and access ever-growing information data bases.

Information technology is an extremely valuable tool, an enabler that is finally making the productivity gains promised during the 1980s. The companies that have thrived through the use of information technology are those that have used a *systems engineering* approach to integrate the technology with the people, process, and organizational aspects of the business.

> *Systems engineering must be the force that integrates technology with the people, product, and organizational aspects of the business.*

Successful organizations in the year 2000 and beyond must be multidisciplined and flexible. They must focus on developing innovative product and process improvements throughout their own business and along the entire supplier-producer-customer enterprise.

These product and process improvements must contribute to improved customer satisfaction and address all stakeholders' needs by continuously improving the company's competitive position in quality, cost, timeliness, product function, and ease of use—as mandated by the customer.

A SYSTEMS APPROACH TO PRODUCTIVITY

Systems engineering, by design, focuses on integrating several major resources. The first is *people*—the work force resource. People improvement begins in schools from kindergarten through the 12th grade and continues through a life-long process of learning.

Next, there must be a focus on *customer wants and needs*. Two types of customers must be served. *External* customers are the primary reasons for being in business. They are the source of the money that keeps the new products coming and the manufacturing machine running. Without external customers, a business will cease to exist. The second customer is the *internal* one. Internally, every employee, every department, every division is a supplier or customer of another employee, department, and division, and attention must be given to the wants and needs of each.

Quality function deployment (QFD), competitive analysis and assessment, and other benchmarking techniques are tools that help capture and satisfy customer wants and needs. Making the customer (internal and external) the focus of all endeavors will ensure that improvement activities will meet the wants, needs, and goals of the business.

> *Customer satisfaction will ultimately determine the success or failure of a company in the normal course of business.*

Many diverse *stakeholders* have a direct impact on a company's success. In addition to the customer, key stakeholders include operations workers, management, and various staff in sales and marketing, finance, and service. Labor unions, distributors, production part and process equipment suppliers, service and nonproduction part suppliers, universities, and local businesses are all part of the business enterprise and must have shared business objectives that lead to a win-win situation. For example:

- Marketing, using realistic images of the product concept, obtains information from the customer as to likes and dislikes long before prototypes need to be built.

- Finance analyzes the concept and works with the producer to determine accurate costs and cost drivers.

- Customers' inputs surface in the form of wants—value, service, and satisfaction—not only in selecting the product, but throughout the *entire* lifetime experience of using the product.

- Service investigates the ease with which the concept can be serviced. The customer is directly affected (as is return business) if the product needs servicing, and quick and successful repair results in a things-gone-right attitude.

- Production workers have a tremendous stake in the success of the company and are usually closest to the product. Their input can significantly improve the process of manufacturing and assembling the product.

- Product and process suppliers will quickly validate their products and processes and meet cost targets.

- Academia can now see many of their intellectual properties brought to market more quickly.

As a company continues to reinvent itself, *up-front disciplines* play a major role in continuous improvement of a product or process. Computer-aided engineering (CAE), simulation, design of experiments (DOE), and failure mode and effects analysis (FMEA) focus on getting at the root causes of things that can go wrong in the process, and put in place the remedies that eliminate or minimize them.

> *Too many companies focus on addressing the symptoms of problems instead of getting at their root causes.*

Once the root causes of problems are identified and understood, a *total systems analysis* approach must be used to develop better ways to execute the process. Too often, a company will optimize a subprocess or element of a process without understanding the systemic cause-and-effect relationship of the entire process.

This "find-it-and-fix-it" approach to quality management started in the post-war era and continued through the 1980s. Today, many companies are taking a total quality management (TQM) approach, using systems engineering to get at the root causes of problems. This structured approach invariably leads to a much improved and more robust process that results in higher quality, lower cost, and reduced cycle time.

Many companies involved in the manufacture of an engineered product have embarked upon programs of re-engineering their business processes as the next panacea for competitive pressure. This grasping of something new is the "Baskin-Robbins" approach to competitiveness—a "flavor of the month" approach.

> *All companies are faced with a change imperative, whether change is driven internally by a sense of discontent or externally by competition.*

The 1990s show that companies embracing change prosper and those avoiding change suffer. Witness the continuing struggles of such great companies as IBM, Sears, General Motors, and K-Mart to reinvent themselves after resisting change for decades. Conversely, look at Chrysler, Hewlett-Packard, Microsoft, Intel, Ford, and Wal-Mart, where ongoing change management has been the key to success.

It is imperative that U.S. industry returns to the basics of people, process, and technology and applies them systematically to the improvement of engineered products. It is a universal theme in American manufacturing circles that the erosion of competitiveness of U.S. companies derives from a lack of vision, short-term focus, negligence in planning, and a failure to effectively implement corporate strategy throughout the enterprise at the shop-floor level. These shortfalls, in turn, can be traced to an almost total lack of attention given to the concept of functional integration of the resources of the organization. Many companies have good people, excellent product and process technologies, and certainly an external push from markets and competition to do better. The key is to step back and take a systems engineering approach to integrating people, product, processes, and technology.

Systems engineering provides a global view of the entire process. Rather than looking at each activity as an independent event, the emphasis of the information age is on examining a process as a total, single system and optimizing the entire system through integration, as opposed to perfecting any single component.

This philosophy engenders a drastic change in our current culture and is fundamental to dramatic increases in performance. Productivity increases can be 10 times, 100 times, and even 1000 times rather than the 10 to 50 percent improvements we are satisfied with today. The integration of people, process, and technology will drive this productivity improvement.

> *A systems viewpoint looks at and optimizes the total process, as opposed to optimizing each individual element in the process.*

The product development process has long been the sacred and private domain of engineers. Management is just now starting to ask the tough questions as to how effectively this process supports time-to-market, and most importantly, time-to-money for new products.

The Elements of Change

TIME AS A CHANGE AGENT

The 1990s will be remembered as a decade of radical change for companies that manufacture engineered products. Product life cycles have become shorter, particularly in the computer industry, to the point that some product designs are virtually obsolete when the product is introduced to the market.

> *Computer-aided engineering is the application of computers to evaluate designs of components, end-products, systems, processes, tools, and facilities.*

Time-based product development, bolstered by a revolution in computer-aided engineering (CAE) tools and processes, has become a strategic imperative. With the recognized gains in market share and profitability from being first to market, companies are developing innovative ways of getting new products designed faster. Engineers, designers, and support teams are becoming increasingly time-driven in their pursuit to achieve customer satisfaction. Bold new initiatives have been implemented to continuously improve the quality and reduce the cost of their company's products.

The Desktop Workstation

The newest addition to the technical toolbox is the engineering workstation. It has become the primary tool of the trade for engineers, designers, and analysts and is credited with boosting engineering productivity by at least 50 percent. But engineering workstation technology is only one element of the success equation. Other factors critical to successful new product development are people, processes, and the organizational and informational structure enabling their interaction.

The $12 Million Engineer

In a typical Fortune 500 company of 1997, the cost of a product development engineer (salaries, fringes, facilities, and tools) is in excess of $400,000 annually. Taken over the normal course of a 30-year career, the lifetime cost of a single engineer is $12 million. This is a staggering investment for any company. As a result, companies must find new ways to maximize the productivity of their product engineers. Even more important than the $12 million lifetime cost, the engineer is the bottleneck for the flow of new product to market. Doubling the productivity of this scarce resource can multiply the revenue and profit stream of the enterprise, providing an important competitive advantage.

> *The fully-burdened cost of a typical product development engineer is in excess of $400,000 annually.*

Process and Mindset Overhaul

Another critical success factor is the product development process itself. Many companies still operate as they did 20 to 40 years ago, and have simply automated their outdated design processes, often with technology that is under-optimized because of lack of planning and knowledge. The long-held tradition in which engineers complete a design and "toss it over the wall" to manufacturing is as dead as the companies that persisted in doing it. Companies looking to survive and thrive must break with the past and carefully plan for, select, create, and nurture. They must identify and establish best practices to re-engineer the product design and manufacturing process.

Continuous People Improvement

To be truly effective, any re-engineered process for new product development must embrace a systems engineering approach that integrates the people, process, and technology. The best and brightest people must be recruited and enrolled in an ongoing training program spanning their entire careers. Rapid changes in business practice and technology growth erode the knowledge of a typical engineering graduate by 50 percent in just 5 years.

Recognizing the critical importance of worker skills, leading-edge companies such as Toyota establish life-long training programs for their engineers. Toyota views people as assets, and like capital assets, the skills of engineers depreciate over time. Toyota has made no secret of its amazement (and amusement) at how typical U.S. companies make hiring decisions. In America, companies will spend weeks, or even months, agonizing over a $1 million machine tool or computer system purchase, but will often make a hiring decision in only hours or days. Toyota views people as a $1 million decision, and goes through an extensive analysis process that weighs teamwork and people-skills as well as technical qualifications.

FLEXIBILITY TO ACCOMMODATE CHANGE

It is no secret that computer-based technology is changing so fast that it has exceeded our abilities to adequately investigate, acquire, and deploy it before its useful life has eroded.

Engineering workstation technology, for example, is changing so rapidly that its optimum life is now less than one year. The ability to quickly upgrade engineering workstation hardware is limited only by application software and more often, by a company's slow internal order fulfillment processes.

The inability of companies to change their product development processes and establish best practices is probably their biggest impediment to competitiveness. Though people, in general, do not like to change, change is arguably the strongest agent driving competitiveness today. A good example of time-to-market and change management is that demonstrated by the personal computer industry. This industry is currently undergoing a major competitive shakeout in the value chain. A Taiwanese company, Mitac, has coined the phrase the "three-six-one cycle," which means 3 months to design a product, 6 months to sell it, and a final month to clear out the inventory before starting over again.

> *In the consumer electronics market, new products that are introduced late often miss more than a third of their potential life-cycle sales and profits.*

Automotive companies have goals of 4-year product cycles, with time-to-market goals of 2 years or less. As one of the largest users of computer-aided technologies, the automotive industry is helping to change the way products are developed. The investment for completely new product development in the automotive industry is large, as much as $3 billion for a new model, and more than $1 billion for a freshened model. About one-third of this investment for new product development is for people costs. Another one-third is for facilities and tools. The remaining one-third is for physical prototypes and testing used to validate the design of the new vehicle.

CATEGORIZING BENEFITS

The benefits of change in the product development process for companies that manufacture engineered products are tremendous. But to fully understand benefits, they first need to be categorized by timing and focus. Table 2-1 shows how the benefits of change will be explained throughout the remainder of this chapter.

TABLE 2-1
Categorizing Change for Maximum Benefit

Change focus	Timing Operational (1-2) years	Strategic (3-5 years)
Business drivers	• Product cycle plan • Competition	• Product portfolio • Market share
Technology drivers	• End-user demands • Installed asset base 　– People 　– Processes 　– Tools	• End-user needs • Envisioned asset base 　– People 　– Processes 　– Tools
Benefits of change	• On-time delivery • Quality management • Cost driver analysis • Customer satisfaction	• Time-to-market • Quality leadership • Cost leadership • Customer loyalty
Enablers of change	• Physical prototypes • Visual engineering • Task orientation • Component focus • Training • Collocated teams	• Mathematical prototypes • Predictive engineering • Process orientation • Systems focus • Core competencies • Virtual teams

The key to realizing the time-based benefit in the product development process is the integration of the design engineer, designer, analyst, and manufacturing engineer into the *virtual* engineer.

Business Drivers

Product cycle plan. All companies that manufacture engineered products have a product cycle plan that is driven by market pull. For example, a typical global automotive company has a product cycle plan that projects more than 10 to 15 new or modified vehicle development programs over a 10-year period. Figure 2-1 illustrates a typical product cycle plan.

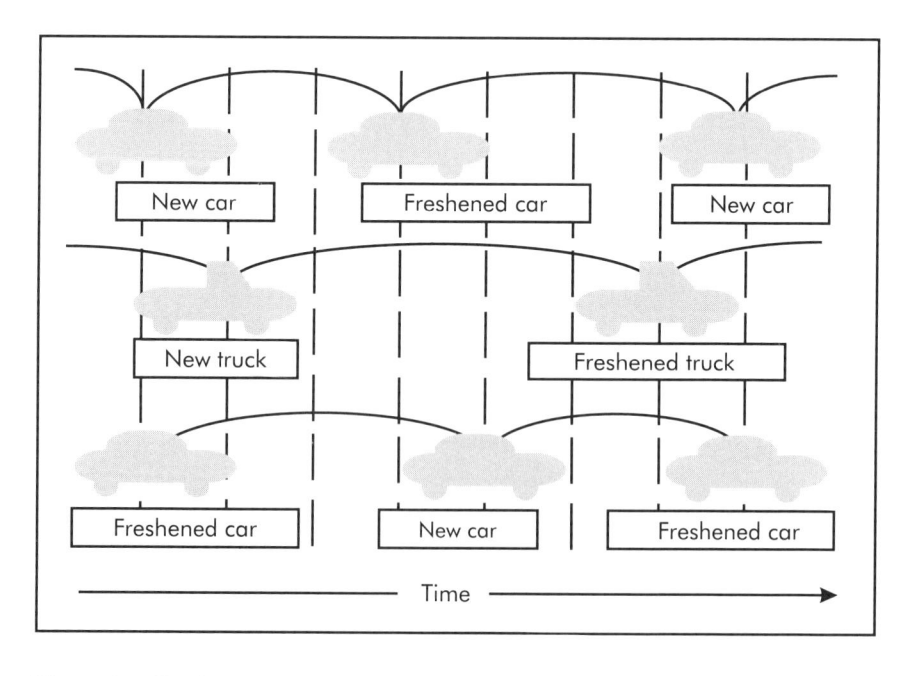

Figure 2-1. Product cycle plan driven by market forces.

These programs will cost $20-30 billion over the next decade. Included in this investment are the people, processes, and tools that will be required to bring these products to market successfully and on time. Product portfolio management ensures that each product life cycle includes the flexibility to "freshen" each product to meet customer and competitive demands.

Competition. The second business driver is competition. Competitive benchmarking became a cottage industry in the 1990s as companies became much more aware that product life cycles were becoming shorter. A company's strategic planning process must be based on knowledge of markets, customers, and competitive threats, including current and anticipated competitors and suppliers. New material technology, and environmental and regulatory threats also impact a company's product strategy. The best companies must look outside their industries and beyond conventional boundaries.

> *In many companies, competition comes from within. It is shrouded in such phrases as "not invented here," "complacency," "lack of focus," and "reluctance to change."*

The product development process can be benchmarked against the competition through a series of time- and cost-based measures that can provide tremendous insight. Some of these measures include:

- Time-to-market or concept-to-customer (months).
- Engineering costs per unit (dollars).
- Warranty costs per unit (dollars).
- Manufacturing costs per unit (dollars).
- Physical prototype/test cost per unit (dollars).
- Mathematical prototype/test cost per unit (dollars).
- Mathematical/physical prototype's time to optimize (months).

While some data may be difficult to quantify precisely, structured analysis can provide surprisingly good estimates that are more than adequate to identify major competitive gaps.

Technology Drivers

End-user demand. In the automotive industry, end-user CAE requirements have rapidly outgrown existing capacity. Increased requirements for simulating crash, noise-vibration-harshness (NVH), fuel economy, durability, vehicle dynamics, computational fluid dynamics (CFD), analytical powertrain, metal formability, and casting/plastic molding behavior continue to drive CAE demand upward. As the slow but gradual paradigm shift from physical prototypes to mathematical prototypes occurs, demand for CAE technology soars.

> *Solids modeling is the representation of a design concept that is developed using 3-D solid objects such as cylinders, cones, and cubes, instead of lines, circles, and arcs.*

CAE technology demand is driven by the need for increased number, size, and accuracy of models for visualization, automeshing, and optimization, which permits substitution of analytical prototypes for physical ones. Solids modeling has changed the way products are developed and evaluated. The demand for CAE to support this new process for product development will double every year through the year 2010. Continuing performance breakthroughs (about 80 to 100 percent annually) in CAE technology, especially engineering workstations, supercomputers, and the information superhighway, will help accommodate this shift. The ability to quickly solve problems will allow emerging predictive engineering software to be integrated into the mainstream; it is a primary enabler to leapfrog the competition.

Installed base. In the automotive industry, the installed base of people, processes, and tools is undergoing a major renewal effort that is being driven by time-based market competition. With the investment for a newly-developed vehicle platform now measured in the billions, and market forces compressing the product development and launch cycle to less than 3 years, this new time-to-market reality has forever changed the way that vehicles will be developed and tested.

> *Computer technology performance is doubling every 12 months. The ability to incorporate it seamlessly and rapidly into existing processes is a core competency.*

The installed base in computer-based systems is massive, and traditionally used annual updates to replace aging and obsolete technologies that were state-of-the art only a few months before. Traditionally, an engineering workstation's life cycle was 60 months (5 years). Today, the useful life of an engineering workstation, one of the critical tools of the trade for engineers, designers, and analysts, has fallen to 3 years and its optimal use is less than one year. This trend will continue into the 21st century. Price and performance breakthroughs are also occurring in supercomputers and networking (the information superhighway), a trend also likely to continue.

A look into the installed base of human resources is sobering. In several automotive companies, more than 50 percent of the hourly and salaried work force will be retiring during the next decade. This shift will have bittersweet consequences. On one hand, it will enhance acceptance of computer-based technologies by a younger and more educated work force. Offsetting this apparent benefit, however, is that much of the time-honed knowledge, skill, and experience, a form of a company's intellectual property, will be going out the door. Major changes in industry such as this are now the norm and the successful companies of the 21st century will be those that embrace change and accept its inevitability.

Most employees will fear, fight, and resist change. Change creates high levels of uncertainty and forces people to operate outside of their "comfort zone." People must not try to brace for change, but loosen up and become more flexible in their approach to their roles and responsibilities.

> *Changes in the workplace affect everyone and create an atmosphere of uncertainty and fear in most workers.*

Success in any change management environment requires flexibility and adaptability of the people. Change can come in the form of re-engineered work processes, insertion of new technology, shifting of markets, or competition from entirely new quarters. But regardless of its form, it is a process, and as such, can be effectively managed.

The responsibility of change implementation lies with management. Management must implement changes positively and communicate relentlessly with all levels. They must embrace change, even if there are short-term costs to retrain and carry excess people. They must create the incentives to reward successful change and intelligent risk taking.

People react differently to change. People who view change as a positive process of renewal will become the change agents and leaders of tomorrow. Companies that employ those people will be the success stories of tomorrow.

People. Most new engineers hired directly out of universities today are comfortable using an engineering workstation as their primary productivity tool, a major difference from the engineer of a generation ago. But despite the computer literacy of the new engineer, there are still many important skills that he or she lacks, that only time will nurture. Figure 2-2 is a skills ledger listing some of the advantages and disadvantages of the newly-hired engineer. These engineers lack only experience to become an effective part of the product development process. The keys to gaining this experience lie in an ongoing training and mentoring program, and an environment that encourages mutual company-employee loyalty.

Process. American manufacturing is finally realizing that the old, serial product development process is an anachronism in a world of time-based marketplace demands. Today, automotive engineers spend less than 25

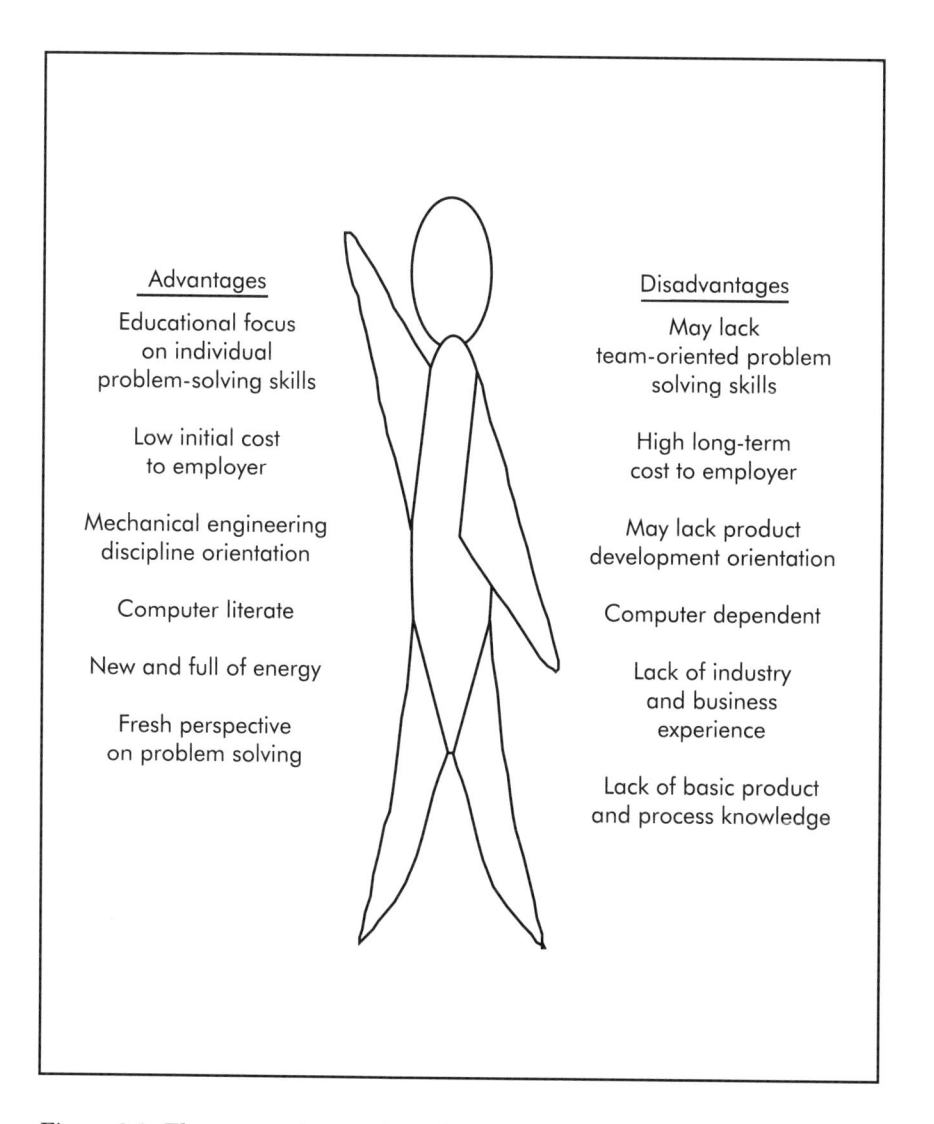

Advantages	Disadvantages
Educational focus on individual problem-solving skills	May lack team-oriented problem solving skills
Low initial cost to employer	High long-term cost to employer
Mechanical engineering discipline orientation	May lack product development orientation
Computer literate	Computer dependent
New and full of energy	Lack of industry and business experience
Fresh perspective on problem solving	Lack of basic product and process knowledge

Figure 2-2. The new product engineer brings to his or her company both benefits and baggage.

percent of their time actually engineering the product, because of the vast knowledge base and the time it takes to search and review it, the long 3-hour meetings to communicate with all affected parties, and the time to administer and manage the support functions.

Figure 2-3 summarizes a typical design process. This process is iterated until the design is optimized or timing constraints force the part to be released. A typical full vehicle process in the auto industry takes from 24 to 36 months.

In today's process, engineering workstations are used for the pre- and post-processing of mathematical models of a

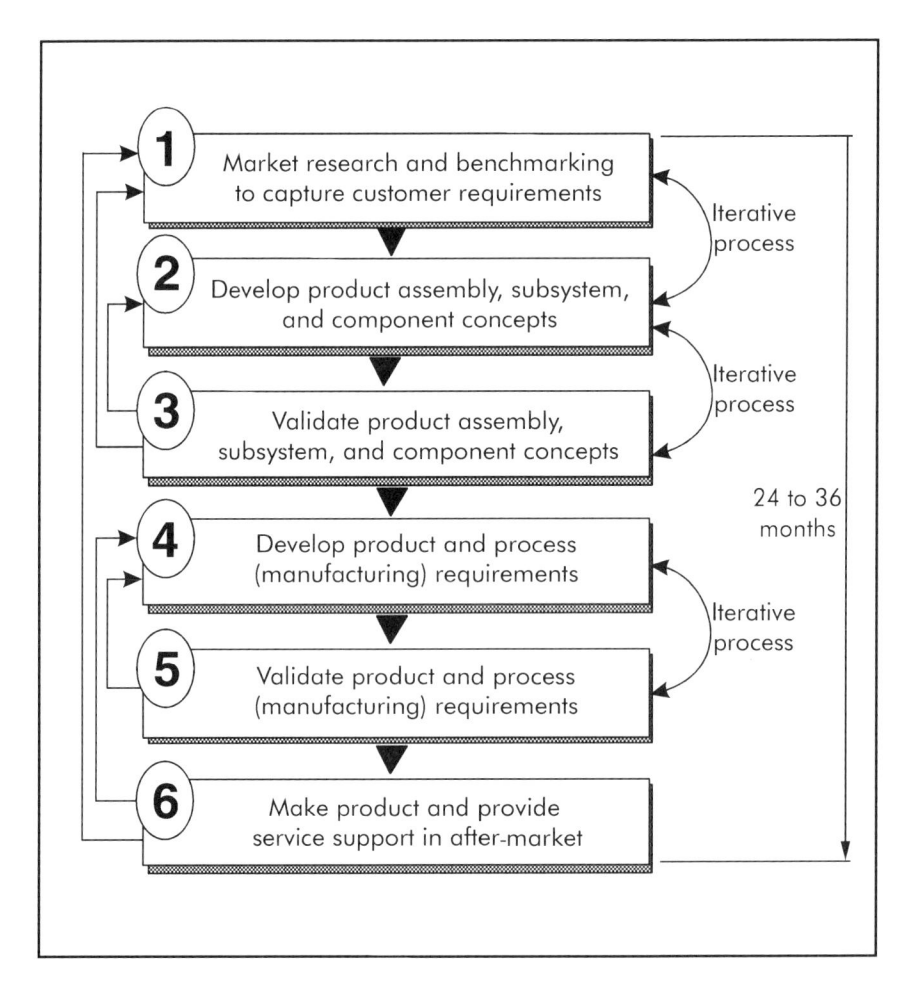

```
1. Market research and benchmarking to capture customer requirements
2. Develop product assembly, subsystem, and component concepts
3. Validate product assembly, subsystem, and component concepts
4. Develop product and process (manufacturing) requirements
5. Validate product and process (manufacturing) requirements
6. Make product and provide service support in after-market
```
(Iterative process between steps; total 24 to 36 months)

Figure 2-3. Current product development practice includes an interative process in which less than 25% of the engineer's time is spent on engineering the product.

design. CAE compute servers, in the form of massively parallel processors (MPPs), scalable parallel processor (SPPs), supercomputers, and high-end engineering workstations are used to analyze, test, and simulate the mathematical models under varying conditions before prototypes are built. Figure 2-4 shows the process graphically. Results from this analysis are passed back to the engineering workstations where the designs are optimized and final details are added. The process of the future will be accomplished on the engineer's virtual desktop. The user is analyzing the full design system response in the form of a virtual prototype, cut from math data and simulated in its climate and boundary conditions. The new designs can be simulated in minutes rather than the days or months currently required. By the year 2010, the simulation of complete systems, such as automobiles, should be performed in 5 minutes.

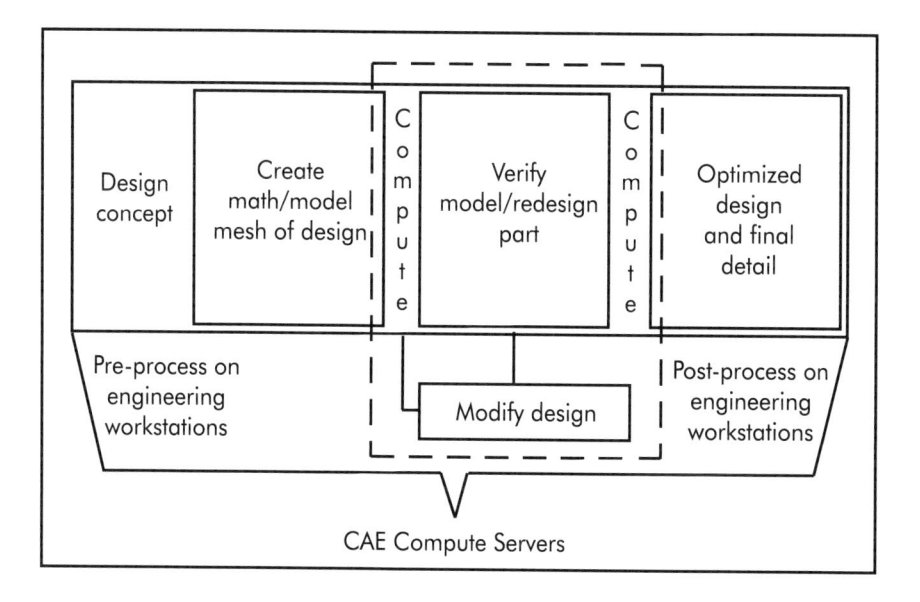

Figure 2-4. Engineering workstations are playing an increasingly important role in product design and engineering, and are rapidly becoming the principal tool of product development.

The technology is in place today to create a *virtual engineering laboratory* on the desktop. This CAE capability will allow construction of even larger models, as well as enable engineers to perform additional verifications and model redesigns before optimization and detail of final design.

Tools. The CAE technology breakthroughs of this decade alone, driven by cost, speed, and power advances in microprocessors, have outpaced all the CAE technology development of the past 40 years. Figure 2-5 projects microprocessor technology advances over the next decade. With continuing improvements in microprocessors, engineers will soon have on their desktops—at a cost of less than $100,000—the equivalent power and performance of supercomputers that cost more than $40 million in the early 1990s.

Technology is a double-edged sword for the computer-minded engineer. Microprocessor capability is increasing faster than the software to run on it. As mentioned earlier, workstation technology performance is doubling every 12 months; software, however, often takes from 18 to 36 months to be rewritten, revised, upgraded, or ported to new hardware.

The new engineer must be prepared and enabled to adapt to the state of the art in both hardware and software development and the increasing capabilities of each. This means training will become a principal part of his or her working routine, one that will be ongoing and focused on continuous improvement. Driving these needs are the constant competitive business demands of introducing higher quality and lower priced products more quickly to increasingly discerning and fickle markets.

> *Technology must be managed in a way that allows hardware, software, and training needs of the engineer to be integrated seamlessly.*

CAE tools available today have evolved from a centralized and time-sharing resource to being readily available on demand on the engineer's desktop. Supercomputers are evolving from dedicated machines to massively parallel and scalable parallel systems.

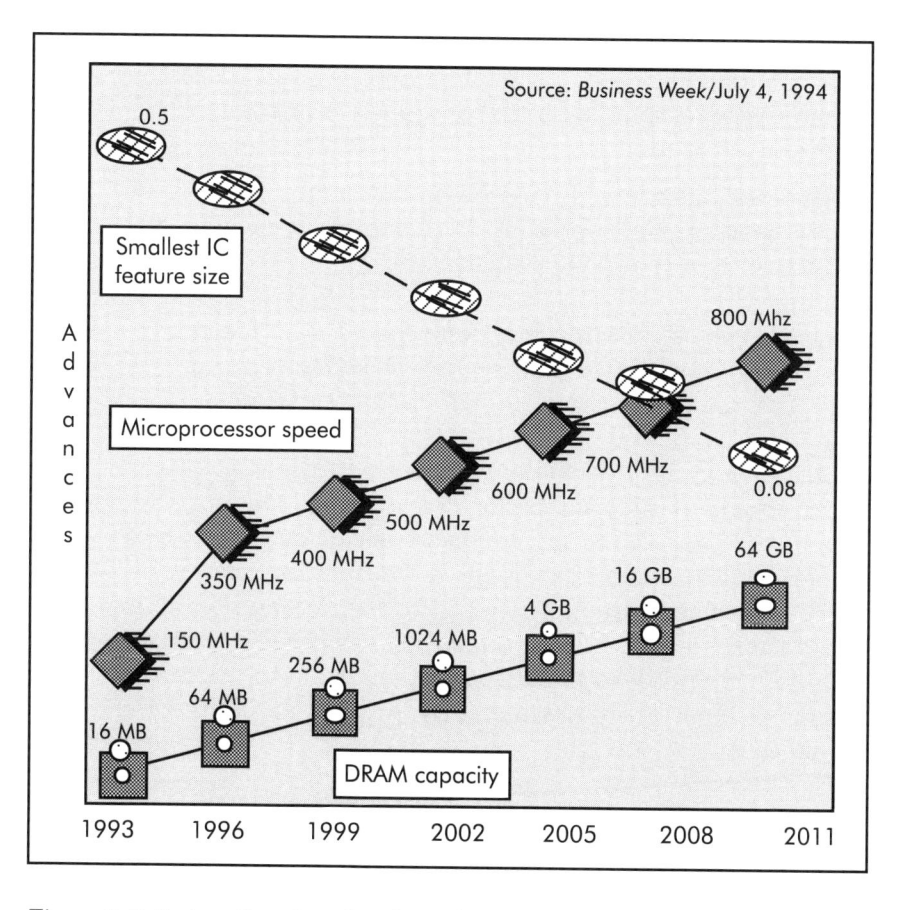

Figure 2-5. In less than two decades, microprocessor technology advances will provide engineers with the power of a supercomputer on their desktops.

Because of its time-intensive nature, software will always be in a catch-up mode to hardware, but continuous breakthroughs in microprocessor performance will allow more robust design features and better performance to be added to software.

Hand-in-hand with improvements in desktop CAE technology are great advances in high-speed communications. Price and performance improvements in fiber optics and wireless technology, and infrastructure development of the world-wide-web, have resulted in unprecedented machine-to-machine dialogue at real-time speeds. These improvements will usher in countless new ways to speed the over-the-line transfer of large files that were previously limited to tapes or mail.

Technology has evolved from being an enabler to a driver as engineers strive to produce higher quality designs at a lower cost, and in much less time.

> *Technology continues to drive the need for faster computer systems, more robust software, and instantaneous communications on the desktop.*

No technology limitations are foreseen in the continuous improvement of microprocessor performance. Microprocessors in the year 2010, at 10 times the capacity of today's models, will become commodities finding application on every imaginable product used in our everyday life. This integration, regardless of product focus, will yield tremendous gains in product quality, value, and time-to-market—and more importantly, it will greatly improve the engineer's productivity.

REALIZING BENEFITS

Many companies today often spend between 25 and 30 percent of their product development budget on physical prototypes and testing. At least one automotive company in 1994 physically constructed about 95 percent of its prototypes—only 5 percent were created mathematically.

Table 2-2 compares existing versus envisioned business practices for a typical automotive company. The increased use of CAE methods, such as mathematical prototypes and solids modeling, will dramatically reduce product develop-

TABLE 2-2
Current versus Envisioned Business Practices

1995 Business Practice	Envisioned Best Practice
• Major physical prototypes at an average cost of $500K	• Mathematical prototypes at an average cost of $1K
• Physical prototypes make up more than 95% of prototypes	• Physical prototypes make up about 25% of prototypes
• Only 5% of prototypes are mathematical (CAE)	• More than 75% of prototypes are mathematical (CAE)
• More than 30% of new product development budget goes toward prototypes	• Less than 10% of new product development budget goes toward prototypes
• As much as 6 months to build, test, and analyze a major physical prototype	• Less than one day to build, test, and analyze mathematical prototypes

ment time and cost. Product development time is reduced by evaluating a design concept in hours rather than months and exploring all the alternative and extreme conditions, which are impossible to prototype physically. This breakthrough yields a minimum improvement of anywhere from 10 to 1 to more than 1000 to 1. The ability to evaluate concepts quickly and cost-effectively provides management with a distinct *strategic advantage*. It allows product designers to choose from concepts that are already optimized rather than merely ensuring that they have performed the function to schedule. This advantage turns product development into a time-based *competitive advantage*. The results of the process can be electronically stored for future use and for continuous process improvement.

The cost reduction opportunity afforded by increased use of mathematical prototypes is in excess of 20 percent of the overall product development cycle plan investment. For major manufacturing companies, this represents a savings measured in the hundreds of millions of dollars for a typical new product development program.

> *The increased use of robust CAE methods enables quantum improvements in the traditional process of product development.*

In development of its 777 airplane, Boeing petitioned the U.S. government to conduct one of its test flights in the computer rather than build an additional physical prototype. By so doing, Boeing avoided spending hundreds of millions of dollars on a test airplane and was able to pass the savings on through the development cycle.

Boeing is considered by its competition to be one of the leaders in the effective use of CAE technologies and methods for new product development. The 777 is considered a technological marvel and was developed extensively with CAD/CAE modeling and rapid prototyping.

> *Boeing and Ford Motor Company are considered the leading practitioners of CAE technologies today.*

Unfortunately, the global market for new airplanes is somewhat fixed and many large airlines are struggling. Upstart regional airlines and an ongoing price war in busy markets have changed forever the way that airlines conduct business. The continuous upgrading of government regulations concerning noise abatement, security, and upgrades of airport early warning systems with state-of-the-art radar technology is adding to the increased cost of doing business. Boeing, in its design strategy, must take all these variables into account. Increased capabilities in CAE technologies facilitate this.

Ford Motor Company is another leading practitioner of CAE technologies. Ford, with more than 10,000 product designers, engineers, and analysts, has one of the largest installed bases of high-performance engineering computing systems in the world. Only a few research and development agencies like the National Aeronautics and Space Administration (NASA), the Department of Energy, and the Department of Defense have more raw high-performance computers than Ford. Moreover, Ford's overall engineering desktop capability is unmatched.

Strategic Product Development Program Benefits

CAE is a strategic initiative and must drive the envisioned process for integrating the methodologies, tools, and best practices of the engineers, designers, and analysts early in the product development process. Such a strategic product development program will result in:

- *Reduced cycle time*, allowing better decisions on quality, cost, and timing. The effective deployment of CAE technology for mathematical analysis on one automotive company's new vehicle reduced prototype build times by more than 30 weeks. This, in turn, reduced overall time-to-market. The company reduced its time-to-market for new product introduction for a new model by more than 5 months. The increased units sold during this "head start" period contributed more than $20 million to the bottom line.

- Design robustness that will *reduce engineering change requests* by 90 percent over traditional new product development designs. In the auto industry, engineering and warranty costs still exceed $1000 per vehicle on average.

- More mathematical models that will *reduce the number of physical prototypes* traditionally used to validate mathematical models and new designs. One automotive company reduced the number of their major component physical prototypes by 50 percent (from six to three).

- The analytic design model will eventually reduce physical prototypes by 75%.

Reduced cycle time. Faster engineering and analysis processes reduce cycle time and allow the new product development program manager to make better decisions on cost, quality, and timing. The effective deployment of CAE technology is expected to reduce traditional product development time by more than 50 percent, through innovations such as:

- More analysis done by using the computer, while still protecting the continued reduction in time of the product development process.

- Evolution from physical prototypes to mathematical prototypes, significantly reducing time in the process.

- Changes to mathematical prototypes in the computer taking minutes and hours instead of the weeks and months required with the traditional process of manufacturing physical prototypes.

- Reduced cycle time from improved best practices in the product development process, primarily from making CAE a strategic part of the process.

- CAE helping to integrate and reduce the product development time for styling, design, mechanical analysis, and manufacturing engineering.

Reduced engineering change requests. CAE technology will allow development of more robust designs, which will reduce engineering change requests by 90 percent over traditional new product development designs.

> *Engineering changes made to the product while still in a mathematical and conceptual design state are up to 1000 times less costly than changes made after the design is being manufactured.*

- More mathematical iterations will be made early in the conceptual design phase, thus ensuring fewer prototype changes prior to formal design release to manufacturing.

- Changes to mathematical prototypes will be made quickly and cost-effectively (on the engineering workstation on the engineer's desktop) as opposed to the expensive and time-consuming changes required on physical prototypes.

- Engineering changes made upstream in the process, while still in a mathematical and conceptual design state, are up to 1000 times less expensive than changes made after the design is in the manufacturing process stage.

Reduced physical prototypes. CAE technology will allow more mathematical models to be developed and prototyped upstream in the product development process. This will reduce the need for the number of downstream physical prototypes traditionally used to validate mathematical models and new designs. It is estimated that the number of physical prototypes will be reduced by as much as 75 percent and fully replaced by mathematical models by the year 2010.

- The continued price and performance improvements of CAE technology and increased functionality of application software will allow the "virtual prototyping" of many designs in the computer without having to build a physical prototype to validate results.

> *Physical prototypes are increasingly being used to validate computer-based mathematical models of the proposed design concept.*

- Increased robustness and accuracy of geometric models will improve design effectiveness of such advanced manufacturing prototype processes as stereolithography and free-form fabrication.

- Physical prototypes will evolve to the point where it will be standard practice to use them to validate computer-based mathematical models.

Improved product quality. Product quality will improve throughout the product development process, from conceptual design rendering all the way through production release. A new generation of CAE technology is currently being developed, and best practices for its effective deployment and utilization will further reduce product quality costs by 95 percent over today's benchmarks.

- The use of a shared geometric model throughout the product development process provides common data to all having a need to know. Such a model will increase the quality of information flow and mathematical robustness of design from styling through manufacturing.

- Visualization and solids modeling will make alternative product designs easier to evaluate and allow changes to be incorporated early in the process.

- CAE technologies such as automeshing and optimization allow faster and more accurate analysis to be performed on an increased number of mathematical models without increasing time in the process.

- Warranty costs remain a major life cycle cost driver of all new vehicle programs. The increased use of CAE technology early in the new product development process will help iterate the alternative designs. This will improve life cycle quality, performance, and reliability decisions.

SO WHAT!

The inevitable shift from mass production, specialization, and technology push to a flexible, information-driven manufacturing system characterized by market pull, functional integration, and time-sensitive production mandates a re-engineered approach to the challenge of change.

> *Time-based competition is forcing much of the change in business today. Many companies are faced with decisions whose outcomes will determine survival or failure.*

- All companies that manufacture engineered products are facing time-based competition as a business driver.

- Time is money. For the manufacturer, time savings translate into profits of millions of dollars per day added to the bottom line for every day that can be taken out of a new product development cycle.

- Product designers, engineers, and analysts must adopt a time-driven mindset in their work, using engineering workstations as the critical tools of their trade. Absence of these tools negatively impacts their productivity by at least 50 percent.

- A systems engineering perspective must be embraced. This viewpoint integrates the critical elements of people, processes, organization, and technology into a balanced system.

- CAE technologies, driven by microprocessor advances and predictive engineering software, are ushering in a "golden age" of time-based product development that is increasingly computer-based.

3

Technology as a Tool

MICROPROCESSOR TECHNOLOGY

As mentioned in Chapter 1, the engineering workstation has become an enabling tool of the engineer. Engineering workstation technology is as fascinating as the technology breakthroughs that have made the workstation possible. At the heart of this technology is the microprocessor, the DNA of engineering workstations. Advances in microprocessor technology have brought dramatic changes to the very concept of manufacturing and its product, process, and growth potential.

> *Microprocessors are the DNA of the engineering workstation, the primary productivity tool of the engineer, designer, and analyst.*

New microprocessor technology can improve engineering workstation price and performance by more than 50 percent annually. This raw increase in performance on the engineer's desktop increases his or her capability exponentially. Of the many questions advanced by management in their quest to boost productivity is one that arises with remarkable frequency: "How can this rapidly growing set of microprocessor changes help me create a process that allows the highest quality products to be developed in the shortest period of time and at a reduced cost?"

To answer the question, it is necessary to understand the rise and pervasiveness of computing power. Microprocessor technology is 40 years old, but has changed more in the last 5 years than in the previous 35. A microprocessor is a single chip that contains all of the computer functions of a central processing unit, making it virtually a computer on a slice of silicon. Just about every computer system or entertainment system today contains a microprocessor.

Microprocessor technology has progressed dramatically over the years since it was first developed in 1947 by William Shockley, John Bardeen, and Walter Brattain, a trio of distinguished physicists all working at Bell Laboratories. Their work completed a long journey of analysis and experimentation that started with the vacuum tubes developed by John Fleming, Robert von Lieben, and Dr. Lee DeForest between 1904 and 1906.

Ten years after the discovery of transistors at Bell Labs, the U.S. electronics industry was born. In 1957, this fledgling industry produced about 30 million transistor-based devices, resulting in about $70 million in industry sales. Fast forward to 1996, where the U.S.-based electronics industry alone achieved more than $100 billion in sales. Most of these sales were driven by an array of new and dazzling products with life cycles averaging 12 months. Today, several major semiconductor consortia or *keiretsus* control the majority of the microprocessor technology development and manufacturing.

Microprocessors and what they do for us daily are taken for granted by most people. The microprocessors that are standard components in our video cameras are more powerful than those used in our advanced defense systems only a decade ago. The technology continues to move forward at dizzying speed, with a doubling of performance about every 12 months.

The growth in desktop computing systems for the engineer would never have occurred without rapid improvements in microprocessor technology. IBM, Hewlett-Packard, Digital Equipment Company (DEC), Silicon Graphics, Inc. (SGI), Sun Microsystems, Intel, Motorola, Texas Instruments, Advanced Micro Devices, and many others have defined their corporate visions around the growth of microprocessor technology.

> *The microprocessor, although taken for granted, has dramatically changed forever the way we go about our daily lives.*

Microprocessor performance is rated on these factors:

- Central processing unit (CPU).

- Clock speed.

- Word length or size, expressed in a number of bits and bytes.

- Memory capacity, measured as the number of bytes of random access memory (RAM).

Although these factors have continued to grow at breathtaking paces, they are not interconnected, and in fact, have very little in common. Long words, or bytes, increase the efficiency of a computer by allowing the machine to process data in larger data packets.

A 32-bit processor is twice as fast as a 16-bit machine, even if the clock speed remains the same. Most systems today are 32-bit, but are quickly evolving to 64-bit machines, and 128-bit machines are expected on the market by the year 2000.

The size of the information content required to describe the design concept, called data packets, is critical to the successful use of engineering/mathematical simulation. As the model size becomes larger, the number of calculations becomes very large and requires more bits for an accurate calculation. A 32-bit machine may be accurate for spreadsheets and simple calculations, but 64-bit or higher machines are minimum requirements for predictive engineering.

High clock speeds result from the integration of the physical design features that are manufactured into the microprocessor. The clock speed performance is a function of the length of the internal circuits, quality of the transistors, and heat dissipation capability. There is no real association between the increase in clock speeds and increased word byte sizes. Today, personal computers operate at 100 megahertz (MHz) and 32-bit word spaces.

> *The workstation on the desk of the virtual engineer in the year 2000 will be a 500-MHz, 64-bit machine with 1 gigabyte (GB) main memory.*

Most early microprocessor designs were based on an architecture of microprocessor instructions that drove the computer program. Each instruction designed into the microprocessor also had to be manufactured on the chip. Early chip manufacturing technology was not sophisticated enough to allow more than a few instructions for each microprocessor. Eventually, the complex instructions were aided by an interpreter that controlled numerous combinations in the form of a microcode. This type of architecture is called complex instruction set computing (CISC). This technological breakthrough for microprocessors was widely used from about 1970 through the mid-1980s, and achieved average annual improvements of about 30 percent.

Starting about 1985, a new and streamlined architecture was developed called reduced instruction set computing (RISC). This architecture has less than 25 percent of the instructions of a CISC microprocessor, allowing designers to place all computer instructions on the chip.

RISC technology facilitates a processing technique called pipelining, which allows loading of new instructions (at the rate of one instruction per tick of the clock cycle) into the processor while preceding commands are being carried out. For comparison, a CISC processor can accept only about one instruction for every six clock cycles.

> *RISC-based engineering workstations have achieved average annual performance improvements of about 70 percent.*

RISC-based technologies were first introduced in engineering workstations around the late 1980s. Starting in 1994, many personal computers started to apply RISC-based microprocessors. All personal computers will likely be RISC-based by the year 2000. Engineering workstations will most likely evolve to using something called super scalar RISC by the year 2000. Super scalar RISC technology will allow unprecedented computing capabilities through a technique called parallel processing.

Parallel processing is a leap forward in high-performance computing that is changing the way many companies access huge amounts of data from ever larger data bases. Parallel processors—which array tens, hundreds, or even thousands of microprocessors in one box—working in tandem—are beginning to provide many more users with the tools they need to finally perform multiple tasks simultaneously. Parallel processing, until recently, was the computing method of choice available only to "coneheads" and "propeller heads" working around the clock on highly complex problems.

> *Parallel processing is the next paradigm shift for using computer-based technologies to solve large problems instantly and on demand.*

The market for parallel processing is just beginning to grow. At the end of 1996, it was a less than a billion dollar per year business divided among SGI/Cray, IBM, HP/Convex, DEC, and a few others. Analysts project the market to grow to almost $5 billion in 1998, with most of this growth coming from commercial users, as opposed to the traditional technical users. Parallel processing users will shift from computational engines and compute servers to a variety of applications supporting product data management, data base assessment, and file serving.

Semiconductor Keiretsus

Keiretsu is a Japanese term describing a conglomerate, or group of independent companies bonded together by a common founding company or by historical linkages. In Japan, there may be cross-holding or the financial institution may hold modest equity stakes in many members. Today, the players are no longer just Japanese and the model has changed with U.S. input. The Japanese model is bound by history, tradition, honor, and equity. The U.S. version is fragile and full of members ready to attack other members—à la Intel or Microsoft. In the electronics industry the equity stakes are less important than the symbiotic relationship that has evolved. They work together to achieve a common goal, and each operates within a specific niche. Figure 3-1 shows a keiretsu comprised of a semiconductor manufacturer, a system manufacturer, a software company, a hardware company, and a distribution company.

All keiretsu members work at their core competencies while pursuing the common goal of all of its members. In the semiconductor industry, essentially five keiretsus exist in the world today.

IBM, Motorola, and Apple Microprocessor Keiretsu. IBM has been very aggressive with the Power PC architecture and its keiretsu with Motorola and Apple Computer. These three have sales volume of at least several million units annually, thus justifying the massive investments required to maintain this architecture. This keiretsu joins the number one company in the world, IBM, with the number two semiconductor company, Motorola as shown in Figure 3-2. To this keiretsu, Apple provides its user-friendly,

Technology as a Tool 43

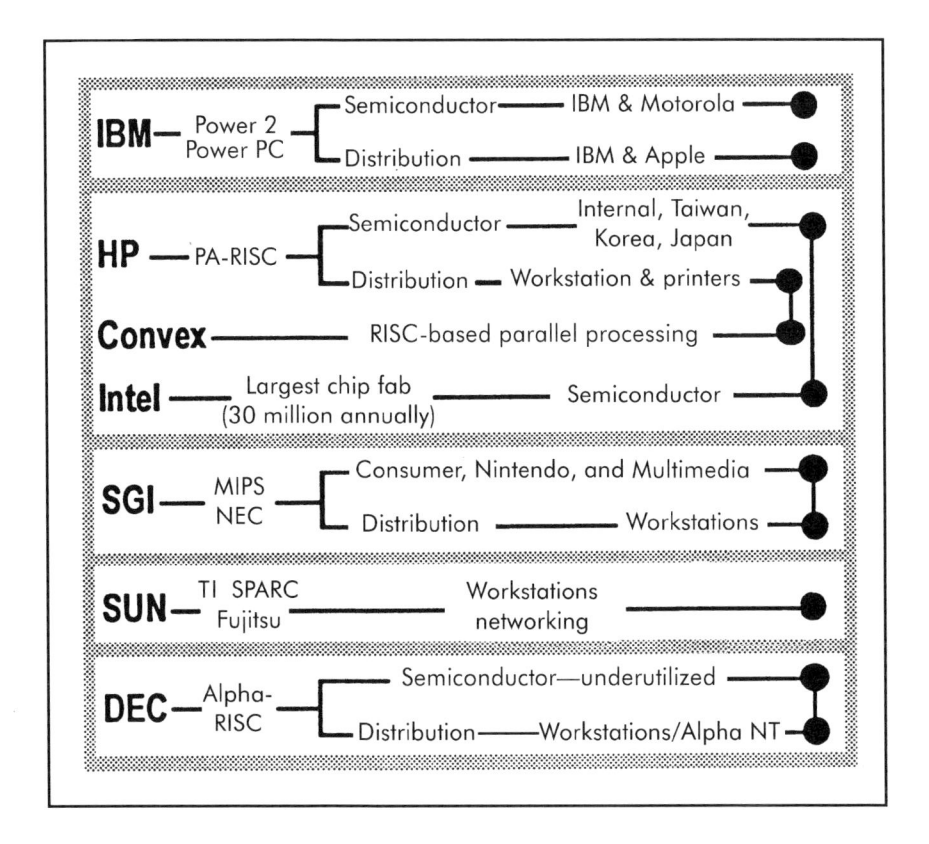

```
IBM ─ Power 2      ┌─ Semiconductor ─── IBM & Motorola ●
      Power PC     └─ Distribution ───── IBM & Apple ●

HP ── PA-RISC      ┌─ Semiconductor ─── Internal, Taiwan, Korea, Japan ●
                   └─ Distribution ── Workstation & printers ●

Convex ─────────── RISC-based parallel processing ●

Intel ── Largest chip fab ──────── Semiconductor ●
         (30 million annually)

SGI ── MIPS        ┌─ Consumer, Nintendo, and Multimedia ●
       NEC         └─ Distribution ──────── Workstations ●

SUN ── TI SPARC    ──── Workstations networking ●
       Fujitsu

DEC ── Alpha-      ┌─ Semiconductor—underutilized ●
       RISC        └─ Distribution ──── Workstations/Alpha NT ●
```

Figure 3-1. Industry collaboration in the form of keiretsus targets specific technologies for development.

object-oriented operating systems that have made Macintosh users dedicated fanatics about the products and their ease of use.

The keiretsu of IBM, Motorola, and Apple establishes a huge consortium whose collective revenues approach $100 billion annually. This consortium has the semiconductor volume and distribution to remain one of the major players. Table 3-1 highlights the keiretsu contributions.

44 *The Virtual Engineer*

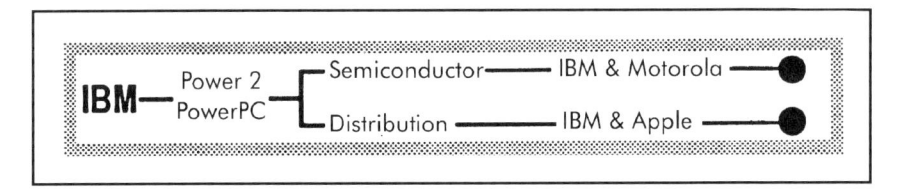

```
IBM — Power 2 / PowerPC ┬ Semiconductor ── IBM & Motorola
                        └ Distribution ──── IBM & Apple
```

Figure 3-2. Each member of a keiretsu brings specific core competencies to the technology being developed.

TABLE 3-1
Strengths of Keiretsu Members

Company	Keiretsu Contributions
IBM	• Number one computer company in the world—committed to Power PC RISC-based architecture to provide scalability from desktop to data center. • Major fabricator and user of RISC-based systems. • Wide array of computer-based systems.
Motorola	• Number two semiconductor company in the world. A major fabricator and user of semiconductors for a wide variety of commercial and technical products. • World-class manufacturer of consumer products.
Apple	• Large user of Power PC for Macintosh systems—owner of famous object-oriented operating system. • Leader in development of graphics and multimedia.

The Power PC RISC-based architecture has become one of the industry's de facto standards, a formidable competitive advantage. The combination of a Macintosh operating system with IBM machines is very appealing and recent Apple/ Microsoft partnering agreements will make their graphical user interface (GUI) the de facto standard for all desktop systems.

HP, Intel, and Convex Microprocessor Keiretsu. Hewlett-Packard Company (HP) has been one of the most successful computer companies of the decade. Most of their success can be traced to a commitment to open systems and not proprietary systems. This commitment along with the acquisition of Convex and an alliance with Intel, has allowed it to become the second largest computer company behind only IBM (see Figure 3-3).

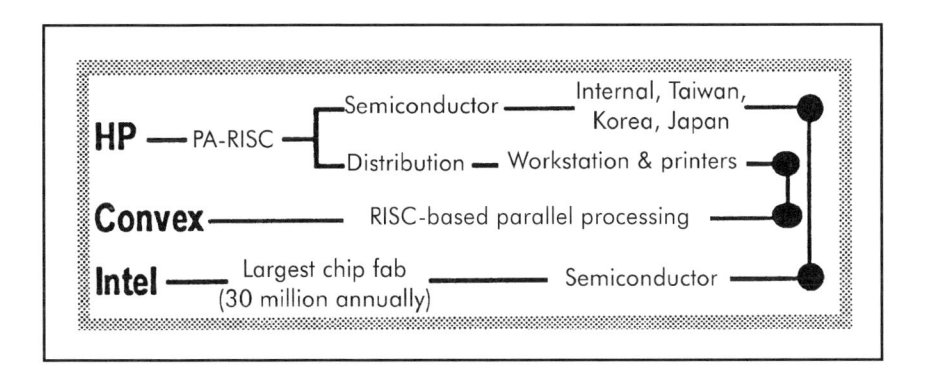

```
HP ── PA-RISC ──┬─ Semiconductor ── Internal, Taiwan, Korea, Japan ──●
                └─ Distribution ── Workstation & printers ──●
Convex ────────── RISC-based parallel processing ──●
Intel ── Largest chip fab (30 million annually) ── Semiconductor ──●
```

Figure 3-3. HP's strategic alliance with Intel has given the company the computing hardware and horsepower to become second only to IBM.

Intel has been another successful company that has become the benchmark by which all other semiconductor manufacturers are measured. With the desktop PC headed to RISC-based processing, Intel needed a major partner and customer, such as HP (and Convex), to migrate the Pentium from CISC to RISC.

The keiretsu of HP, Intel, and Convex comprises a consortium with assets approaching $50 billion. This consortium, outlined in Table 3-2, has the HP and Intel semiconductor volume and HP's printer distribution channels.

TABLE 3-2
Member Strengths of the HP Keiretsu

Company	Keiretsu Contributions
HP	• Number two computer company in the world—committed to RISC-based architecture to provide scalability from desktop to data center. • Major fabricator and user of RISC-based systems. • Dominates global market for inkjet and laser printers.
Intel	• Number one semiconductor company in the world with fabrication capability of over 30 million units annually. • Will merge its CISC-based Pentium suite of products into a next-generation RISC design with Hewlett-Packard.
Convex	• Developer of large, RISC-based, high performance computing systems that are scalable from desktop to data center. • Leader in development of parallel processing enablers. Acquired by HP in 1995 as minisupercomputer division.

The HP/Intel/Convex RISC-based computing architecture will likely become an industry standard. The merging of RISC, Pentium, and super scalar parallel processing on the engineer's desktop will provide powerful tools for the virtual engineer.

SGI, NEC, and Nintendo Microprocessor Keiretsu. SGI has successfully made the transition from an entrepreneurial start-up company to a full-fledged computer graphics company. SGI is the computer graphics industry leader (see Figure 3-4).

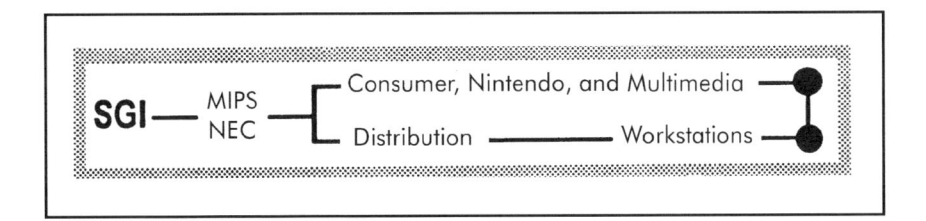

```
┌─────────────────────────────────────────────────────────┐
│         MIPS   ┌─ Consumer, Nintendo, and Multimedia ─●│
│  SGI ─        ─┤                                       │
│         NEC    └─ Distribution ──────── Workstations ─●│
└─────────────────────────────────────────────────────────┘
```

Figure 3-4. SGI, the industry leader in computer graphics, has teamed with NEC and Nintendo to form a keiretsu with annual revenues of nearly $50 billion.

SGI acquired the Cray Company in 1995. Nippon Electric Company (NEC) also produces some traditional, vector-based, supercomputers. The SGI (Cray)/NEC keiretsu approaches $50 billion in annual sales.

SGI (Cray), NEC, and Nintendo together provide the production volume. SGI has a strong semiconductor partner as long as NEC continues to make massive investments in new semiconductor technology and fabrication facilities. Table 3-3 highlights the keiretsu member contributions.

After studying the IBM- and HP-based keiretsus, the logical question is, "Is there really a place for a third, fourth, or fifth keiretsu?" SGI, NEC, and Nintendo have the volume and technology and should survive the shakeout.

SUN, TI, and Fujitsu Microprocessor Keiretsu. The remaining keiretsu player is Sun Microsystems, Inc. (SUN). SUN does not have affiliations that yield the volumes of the other keiretsus (see Figure 3-5). It is not clear whether Texas Instruments (TI) or Fujitsu will continue with the SPARC architecture with its low volumes.

SUN, TI, and Fujitsu combine for more than $50 billion in revenues annually. SUN's strength is in its retail and commercial workstation market, but although recognized as one of the installed-base workstation market leaders, it is losing out in the race with HP, IBM, and SGI for preeminence on the engineer's desktop.

TABLE 3-3
SGI Member Contributions

Company	Keiretsu Contributions
SGI	• Owner of some of the best price/performance workstations in the business, especially with its dazzling graphics. • Acquired major competitor, Cray, thus gaining a large and impressive installed customer base globally. • Does not make its own chips. Workstation volume alone is not enough to justify new investments in chip technology.
NEC	• Key semiconductor player in keiretsu. Has world-class facilities and all important entry into closed Japanese market. • Cannot effectively compete against IBM- and HP-based keiretsus. Not sure if industry can support three standards.
Nintendo	• One of the most profitable companies in the world. Holds the key to establishing low-cost distribution to retail markets.

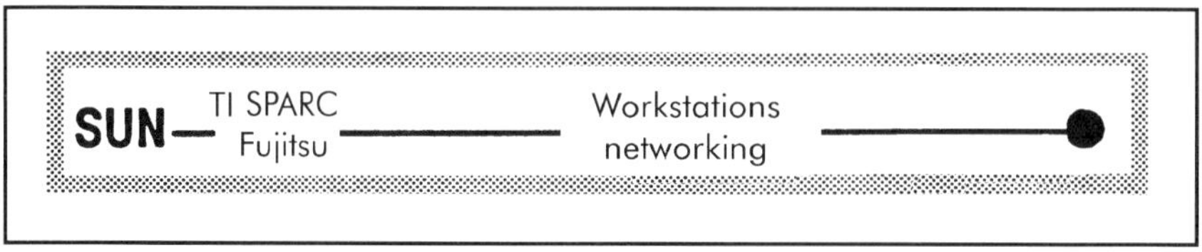

Figure 3-5. Another player in the workstation market is the SUN, TI, Fujitsu keiretsu.

The keiretsu of SUN, TI, and Fujitsu, outlined in Table 3-4, provides volume, but the question is whether its architecture can survive against IBM, HP, and SGI. TI is a

TABLE 3-4
SUN, TI, Fujitsu Keiretsu

Company	Contributions
SUN	• One of the first workstation companies focused totally on the commercial and technical desktop market. • Evolving to JAVA-based and web-based computers as next paradigm of computing as alternative to WINTEL. • Does not make its own chips. Workstation volume alone is not enough to justify any investments in chip technology.
TI	• Key semiconductor player in keiretsu. Has world-class facilities and all-important entry into closed Japanese market. • Cannot effectively compete against IBM- and HP-based keiretsus. History of teaming with Japanese companies.
Fujitsu	• Japan's clone of IBM. Large manufacturer and user of CISC-based semiconductors—must move to RISC.

world-class manufacturer of semiconductors for a wide variety of commercial applications. Fujitsu is a Japanese company that has cloned much of IBM's success in the mainframe and CISC-based market. The question facing both TI and Fujitsu is whether they move on to another RISC-based architecture or remain committed to becoming a niche player in the high-stakes semiconductor arena.

Many industry experts predict that only the IBM-led and HP-led keiretsus will survive the upcoming shakeout of the semiconductor industry.

Digital Equipment Company (DEC). DEC is one of the weakest industry players (see Figure 3-6) and really not a keiretsu at all. Its Alpha chip was late to market and though very powerful, has had difficulty breaking into other keiretsus' equipment. It takes the broad relationships of the other keiretsu members to create the Alpha chip's market. Its semiconductor fabrication facilities are underutilized. Advanced Micro Devices (AMD) is fabricating its chips in DEC's fabrication facilities.

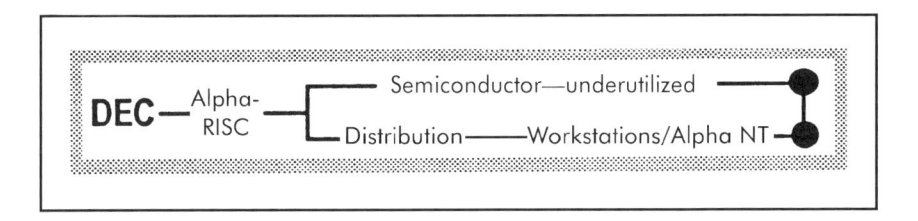

Figure 3-6. DEC, later to market with its latest chip, is a small player in the industry as compared to the other keiretsus.

The Alpha/NT combination also has not been overly successful. Other concerns are DEC's eroded market position and lack of partners which have prevented it from making the needed investments in future semiconductor technology. However, DEC is in litigation with Intel for patent infringements. This legal action may take years. Another scenario could have Intel licensing DEC technology.

DEC, as outlined in Table 3-5, is undergoing fundamental changes in the high-end workstation environment. Once the high-flying company in the 1980s that criticized IBM's proprietary operating system, DEC has also fallen victim to its own success with VAX VMS. Although DEC still gets about 40 percent of its revenues from VAX, it is torn between making its super-fast Alpha RISC chip a standard, while reluctantly moving its installed base from proprietary VMS to open systems.

DEC does not have enough money for the necessary investments to be a major stand-alone RISC player. Many

TABLE 3-5
DEC Offerings to the Information Industry

Company	Contributions
DEC	• Introduced Alpha RISC chip, one of the fastest in the industry, but has been unable to get software ported. • Semiconductor factories are idle. DEC has outsourced production of chips to Advanced Micro Devices in its own underutilized facilities. • May not be able to move quickly enough or have enough money for necessary reinvestment to remain a player.

observers feel: that DEC could end up as a billion dollar integrator of other companies' systems; or it could be an acquisition candidate to bolster other keiretsus; or it could become history as many other famous names of the past in the electronics industry.

SEMICONDUCTOR INVESTMENT DRIVERS

Semiconductor technology advances are driven by feature sizes, dynamic random access memory (DRAM) capacity, clock cycle speed, address space, industrial consortia and keiretsu partnerships, and ability to invest. Feature sizes (in the form of micron line width) will change from an existing 0.5 microns to less than 0.08 microns over the next decade. DRAM capacity is projected to quadruple every three years, resulting in integrated circuits of 256 MB DRAM by the year 2000. New technology that uses light or new materials like copper, as announced by IBM, will solve problems like heat and space and allow speed improvements that far surpass Moore's law—the doubling of computer speed every 12 months.

Volume is the only feasible way to recoup the massive investments required for semiconductor fabrication facilities. Intel has stated that it costs about $1 billion for one of their high-volume fabrication facilities to produce the current 0.5-micron width. Estimates for new facilities to produce 0.35-micron wafers are $2 billion, and $4 billion to achieve 0.25 microns near the end of this decade.

This ability to invest has started an initial shakeout of the semiconductor manufacturers and engineering workstation suppliers. It may be that only two major keiretsus will be strong enough to establish de facto standards: the IBM-led and HP-led keiretsus. Together, they will represent more than 85 percent of the desktop machines.

ENGINEERING WORKSTATION PRICE AND PERFORMANCE

Semiconductor technology advances drive engineering workstation technology. Traditional minicomputer systems were based on complex instruction set computing (CISC) architecture from 1970 to 1985. The introduction of reduced instruction set computing (RISC) architecture in 1985 enabled the emergence of workstations, making them cost-effective engineering tools. Today, super scalar RISC allows annual improvements of around 100 percent. Figure 3-7 reviews the technologies over the past 20 years.

Technology price and performance improvements are making engineering workstations virtually a commodity. For companies to take advantage of rapid technology improvements, systems must meet industry standards and conform to new hardware designs to enable upgrades without disruptive, costly, and time-consuming changes to the working environment. Electronic upgrades and snap-in or programmed-in retrofits will be implemented over the network (information highway) without disrupting the end user.

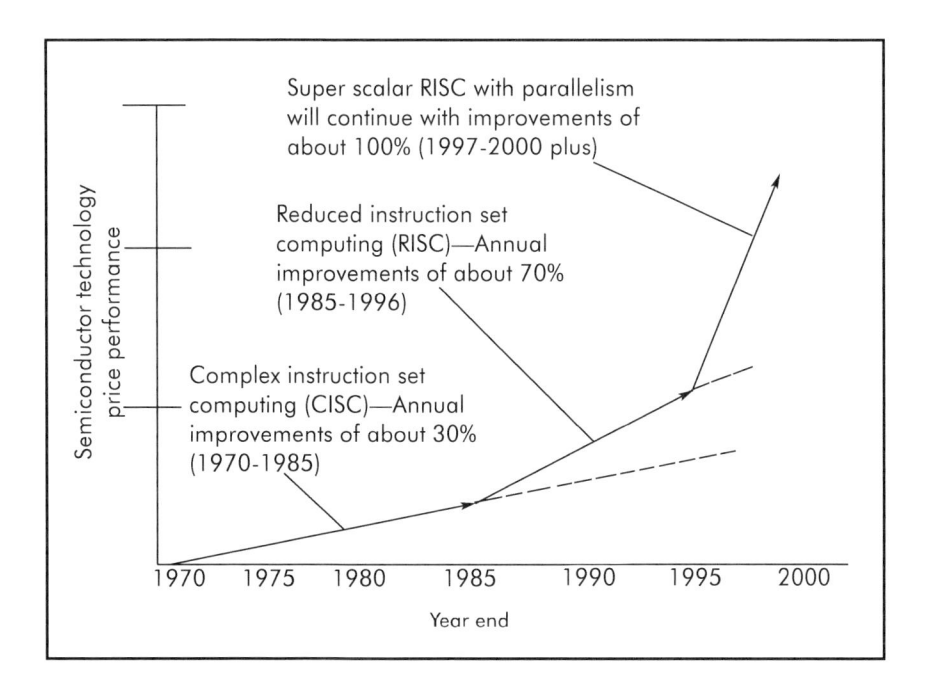

Figure 3-7. Super scalar RISC with parallelism will continue to generate price/performance advances well into the next century.

SOFTWARE

The word "software" has been in widespread use since about 1960, when programs were first sold separately from the computer hardware on which they ran. Software refers to the instructions that tell computers what to do. Without these detailed instructions, a computer is a useless tool.

The creation of software involves three basic steps: defining the problem, devising a plan to solve it, and then translating the solution into a language that computers can understand.

> *The software industry has a long way to go to match the quality, price, and performance of the semiconductor industry.*

Software is often the last element to reach the market since the machine on which it must run first must be developed and tested. Computer-based simulators are often used today to analyze various benchmark programs and predict their performance. The simulators have acceptable error rates, are slow, and difficult to learn.

Though several good operating systems exist, for the purposes of this discussion, we focus on UNIX, a standard operating system preferred by most engineers. Although UNIX represents only about 5 percent of all desktop operating systems, it is the de facto standard for engineering desktops. UNIX is very good at allowing different machines to communicate and perform multitasking functions within a local area network. UNIX's ability to communicate with various machines makes it an ideal candidate for systems integration.

SO WHAT!

Technology as a tool is only as effective as the level of integration it enjoys with human, organizational, and process resources. Implemented as part of a total systems approach to manufacturing, it will produce the strategic competitive advantage necessary for success in the 21st century.

- Technology is advancing so fast that it is impossible for traditional process control to make it work. Management must empower those closest to the changes that are occurring to achieve continuous improvement and seamless adoption.

- Semiconductor technology is doubling in performance every 12 months.

- Software selection will become even more important, and system interoperability will be the key.

- The semiconductor shakeout is underway. Several major industrial consortia or *keiretsus* may well control the technology offered to the marketplace. Or software may break these bonds and emerge as the discriminator. The choice of technology is now affected by predictions about who's going to be around next time.

- Engineers, designers, and analysts will have the equivalent of desktop supercomputers to assist them in the pursuit of high-quality, low-cost solutions.

- The continuing semiconductor price and performance breakthroughs will go a long way toward solving many of today's software performance problems.

- The software development process will change to enable new hardware architectures to leapfrog the existing price/performance.

- Computer-aided engineering technologies will become increasingly affordable to many smaller companies, greatly increasing their market competitiveness.

4

Reinventing the Current Process

INFORMATION AS AN ECONOMIC ASSET

The U.S. enjoys one of the highest living standards in the world today, thanks in large part to manufacturing. Much of the wealth created in America was derived from the manufacture of basic and engineered products—steel, plastics, paper, automobiles, aircraft, heavy equipment, and capital goods. Until recently, more than one out of every four jobs depended on the manufacturing industry.

> *Microsoft creates more derivative jobs in the Seattle area than Boeing, although its annual sales are only a fraction of Boeing's.*

Today, the information systems industry is driving the U.S. economy. Two key global markets are dominated by U.S. companies. On the processing side, Intel Corp. owns 90 percent of the microprocessor market with a product that reinvents itself every 12 months. In the software arena, Microsoft controls 80 percent of the world market, most of it for personal computer applications.

This chapter describes a new business model and an associated set of techniques that executives and managers will need to reinvent the engineering process. This reinvention emphasizes development of the highest quality products in the shortest period of time at costs less than today's, in response to unpredictable, changing market demands.

The changes occurring in most companies are predictable and have roots in the early 1960s. These changes are fundamental and affect virtually every aspect of our living practices. Just as the Industrial Age totally changed the economic and social structure of the world, the Information Age will fundamentally change our new economic and social structure.

> *Henry Ford drastically changed the industrial landscape with automation of the manufacturing process; the microprocessor is now altering today's industrial picture just as profoundly.*

In 1980, the PC, coupled with financial spreadsheet software, changed our entire view about using computers rather than conventional calculation techniques in the everyday workplace. Hardware and software breakthroughs put a time stamp on the Information Age. Where the Industrial Age focused on converting raw material into product, the Information Age is about deploying a vast intellectual base to a wide variety of problems not tied exclusively to the shop floor. This evolution has changed the underlying foundation of most industries.

During the mid-1980s, computing power shifted from centralized mainframe data centers to the desktop. The emergence of ever more powerful engineering workstations (sometimes referred to as high-end PCs), high-speed communications, and minisupercomputers allowed engineering data to be processed and analyzed on the desktop.

Another process, visual engineering, is the ability to represent traditional CAD line drawings as realistic objects. The new representation allows complex engineering drawings to be instantly understood by anyone viewing them. This breakthrough was achieved through microprocessor improvements in speed and cost reductions. Processing time was reduced from 1 week to 2 minutes or less and the cost of the computer from $1 million to less than $40,000.

> *Computing power has migrated from data centers to the desktop and from alphanumeric data to visual engineering data.*

ENTER VISUAL ENGINEERING

These dramatic computer technology breakthroughs allow fundamental changes to occur in the process. Rather than improving an existing process incrementally, the management team now has the capability to leapfrog time-consuming steps and fundamentally change the way design and analysis are performed. This greatly-expanded capability allows a new process—*visual engineering*—to become an integral part of product development. Visual engineering has ushered in an entirely new era in the way engineered products and processes are developed, validated, and put into production.

Visual engineering allows all stakeholders in the product life cycle to see the same image of the concept or drawing—from the design engineer who brings ideas from concept to manufacturing readiness, to the manufacturing engineer who brings designs to manufacturability, to the product service engineer who brings value to the customer. This new phenomenon in the engineering world allows product features to be quickly understood and compared with alternative designs for quick resolution of design conflicts.

With visual engineering, realistic images of an engineered product or process are depicted in a fraction of the time and at a fraction of the cost of building a demonstration prototype. This time and cost savings translates into a significant productivity gain. An additional key capability is the ability to view the component actually functioning within the entire system. This new dynamic capability holds the promise to analytically prototype the entire system before physical prototypes are built.

> *Visual engineering is a major breakthrough in desktop computing technology, allowing all with a need to know to see exactly the same image.*

One of the problems with engineering productivity in the past was the inability to effectively communicate design intent to people in various other functions involved in the process. One had to have some basic drafting and blueprint reading experience to be able to interpret engineering drawings and make any reasonable deduction as to their form, function, and design intent.

Visual engineering has changed all of this so marketing, planners, product development teams, design engineers, manufacturing engineers, serviceability engineers, and suppliers see the same drawing and the same functioning part at the same time. This facilitates common understanding, reveals system interferences, reduces time in the decision process, and precludes anyone in the design loop from working to drawings that are not the most current.

In the past, centralized and mainframe-based data centers were the domain of the information systems (IS) department. Since most information-based applications of 20 years ago were financially oriented, most IS departments were under the management and control of the finance or accounting function of the company. This centralization resulted in power to a select few, and in the development of an elitist organization that was not really end-user (or customer) -driven.

> *Engineering end-users were especially slighted since their requirements were difficult for IS providers to understand.*

Microprocessor technology, giving birth to the PC and high-speed communications, changed this in the 1980s. The emergence of high-end PCs in the form of workstations, and advances in high-speed communications and minisupercomputers created a struggle between IS as providers and controllers of computing services and product development as customers (end-users). The struggle took place on two levels. The marketplace developments in software threatened the traditional internal program code-writing roles of the systems organizations and offered to meet the demands of the engineer for a state-of-the-art workstation. Each threatened the MIS manager's traditional budget and resource control paradigms. The struggles typically add 9 to 12 months to the order process (see Chapter 7, case study three).With time now as a business driver, it is critical that information be rapidly deployed to the desktop in the hands of the end-users, putting computer-driven design freedom where it belongs.

How does Microsoft, with sales of only about $5 billion have the same market value as General Motors, which has sales of more than $150 billion? How does one explain how Nintendo achieved over $5 billion in sales—and over $1.5 billion in profit—with fewer than 1000 employees? Microsoft and Nintendo are in the information technology business.

> *Microsoft and Nintendo are today what General Motors and IBM were in their industry dominance in the 1970s.*

The ability to quickly deploy intellectual capital is important for success in industry. Manufactured products today are the result of information management and deployment and the company with the capability to deploy faster than competitors has the market edge. One automotive company estimated that it costs them more than $4 million in lost profits for each day that a new model is late to market.

The fundamental shift to a time- and knowledge-based process will determine the survival of any company involved in the manufacture of engineered products. The measure of success will be totally different in the 21st century. Just as Henry Ford changed the industrial landscape with automation of the manufacturing process in this century, information technology will drastically change the new engineering and product development process in the next. It will affect the financial, legal, manufacturing, health services, and engineering professions by dramatically increasing not only the speed at which their objectives can be achieved, but the capacity for archiving and recalling records and modeling processes for continuous improvement.

Rather than being distinct disciplines, departments within organizations will begin to lose their individual boundaries and evolve into a highly-integrated system using the emerging information superhighway (intranet/internet) and computer-based technologies as the common catalysts for productivity. This system will work in real time, on demand to the requester, and in a form that is instantly recognizable by the user, regardless of his or her specific discipline. Knowledge will be transferred seamlessly.

But to reinvent their engineering processes, senior management must throw out their current notions about organization and work flow. Rather than looking at each activity independently, the emphasis in the Information Age must be on examining a process as a total, single system and optimizing the total system rather than any single component. This philosophy represents a dramatic change to our current culture and will result in dramatic increases in productivity. Process improvement increases will be 10, 100, and even 1000 times that of the traditional 10 to 50 percent improvements we see today. Pie in the sky? Not really.

Traditional process improvement focuses on small, incremental improvements to the current process.

Many companies have been involved in traditional process improvement, an approach that looks at optimization or improvement of individual activities or steps in the process. This book is not about improving discrete processes, but rather a "step out of the box" challenge to the existing processes—a self-examination that will allow re-engineering of the process from the "voice of the customer" and "voice of the process."

Given the magnitude of change in manufacturing and the intensity of competition, it should be clear to senior management that current processes are no longer adequate and must be changed. The traditional approach to process improvement is to examine the current process and identify those process activities or steps that can be automated using computers. This business strategy works if the sole objective is to go after small, incremental changes. However, for the 21st century, this business strategy will no longer work.

Rather than optimizing the current process, let's step back and look at the primary objective of product development. Management's objective is to produce the highest quality product in the shortest period of time, at a reasonable cost.

Most companies that manufactured engineered products before the 1980s were not customer-driven. Requirements for new products were often dictated by an unenlightened marketing or engineering department. "Customers be damned" was often their motto, and they refused to listen to customer-protective salespeople, weary service people, or less-than-respected manufacturing personnel. The Japanese helped to change this process. Customers are now the product focal point and their tastes are more discerning and their demands more rapidly changing than ever before. Market pull has replaced technology push, and woe be to the manufacturer failing to respond.

WHO'S LISTENING TO THE CUSTOMER?

A real-world story serves to illustrate. Engineers at a dog food supplier developed what, in their opinion, was a world-class product that exceeded all daily nutritional requirements, had a shelf life of years, and was to be produced in a state-of-the-art, low-cost factory.

The problem with the product was that its development took place without input from the customer—in this case, dogs. As it turned out, the dogs hated the product. So much for capturing the "voice of the customer" early in the design process. Such stories pervade the American manufacturing experience.

The voice of the customer is structured feedback using tools such as Quality Function Deployment (QFD). This defines the process steps that will design a product to meet expectations for quality, value, and service.

In the 1980s, companies began to recognize the phenomenon called the "voice of the customer." Manufacturers were finally starting to pay attention to what the customer had to say about his or her wants and needs. In turn, engineers actually used customer feedback and integrated market research and competitive benchmarking into the design process. Compared to many practices before the decade of the 1980s, this was truly a breakthrough.

Using the voice of the customer to incorporate their desired primary and secondary features into the design, while meeting minimum design requirements for function, results in creating exciting product features that meet or exceed customer expectations.

Forward-looking companies are becoming more process-driven. Time is money and waits for no one, and increasingly, time is becoming the core element in defining corporate strategy. Just as total quality management drove companies in the 1980s, time and speed drive them today.

WHO UNDERSTANDS THE PROCESS?

Companies that have implemented Japanese-style, Just-in-Time manufacturing techniques have found the product development process and the entire supply chain to be major areas of opportunity for increased competitiveness. The voice of the process is an effective measure of how long it takes a company to perform a process and how well it performs it.

> *The voice of the process is used to map the inputs, subprocesses, and outputs for analysis.*

Many companies find it valuable to simply flowchart their business processes. Others discover that no single person fully comprehends the entire process, and are surprised at the large number of subtasks, handoffs, and different people required to complete even a simple process.

The first-run capability of any process to perform as expected is usually very low, resulting in much rework or an overabundance of inspectors and checkers in the system. Elapsed time is also revealing. Companies are often shocked when they find out how long it takes to execute seemingly simple administrative process tasks, like ordering an engineering workstation.

THE PRODUCT DEVELOPMENT PROCESS

Clearly, all companies involved in the manufacture of engineered products must have some sort of product development process. The journey from market need through conceptualization to production follows a defined and fairly well-understood road map of tasks. Figure 4-1 depicts the major steps of a typical product development process.

For a new automotive vehicle design, the new product development process typically takes 18 to 24 months, depending upon complexity of design, number of models offered, and the percentage of newly-designed parts versus the number incorporated from previous designs. Let's look at each step of the process in detail.

Step One: Market Research and Benchmarking

For a company to increase its market share, it must have a time-based process for capturing the needs and expectations of its customers. Studies have shown that it is eight times more costly to recapture a customer after he or she is lost than it is to satisfy and retain the customer in the first place; hence, the voice of the customer takes on strategic significance (see Figure 4-1).

Quality function deployment. Step one of the process, market research and benchmarking to capture customer requirements, uses a technique called quality function deployment (QFD).

QFD is a set of techniques that has become extremely popular with companies that manufacture engineered products, particularly consumer products. It combines process elements of total quality management (TQM) to focus cross-functional activities on the voice of the customer. This disciplined, systematic process maintains the integrity of the customer's wants and needs and provides the structure to respond faithfully to the customer's voice.

QFD was created in the early 1970s by Mitsubishi Heavy Industries at the Kobe shipyards in Japan, where ocean-going vessels were built to military specifications. Shipbuilding requires a great deal of capital outlay to produce even one copy of the product. This fact, combined with stringent government regulations that had to be complied with, forced Kobe shipyard's management to commit to some form of effective upstream quality assurance.

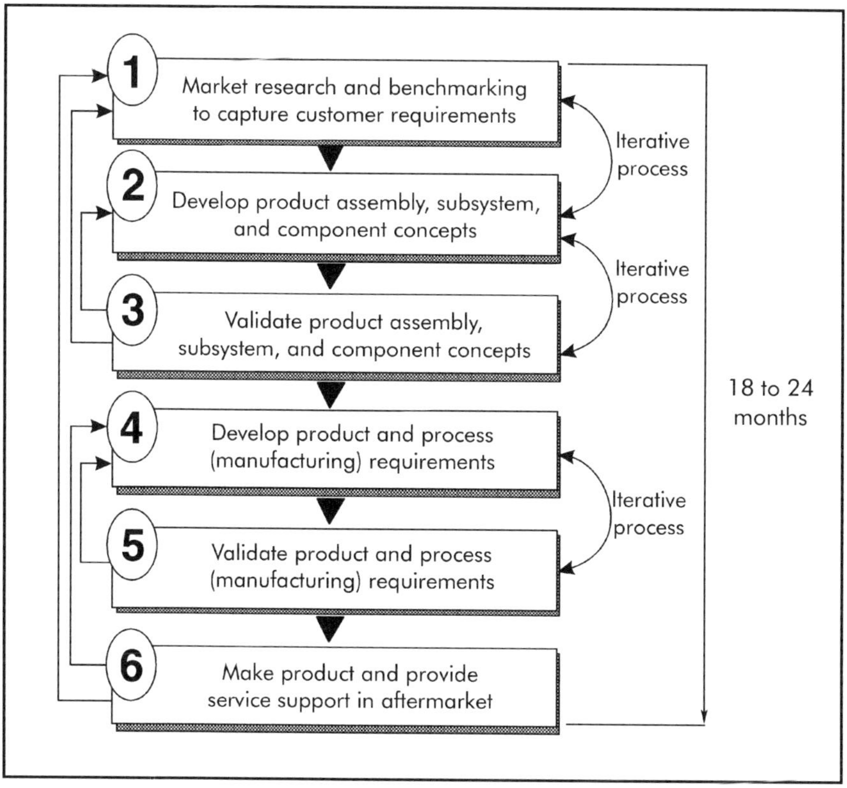

Figure 4-1. Typical product development process.

Flow (1→6) with iterative feedback loops, spanning 18 to 24 months:
1. Market research and benchmarking to capture customer requirements
2. Develop product assembly, subsystem, and component concepts
3. Validate product assembly, subsystem, and component concepts
4. Develop product and process (manufacturing) requirements
5. Validate product and process (manufacturing) requirements
6. Make product and provide service support in aftermarket

> *QFD is a set of techniques created by the Japanese to convert customer requirements into design and manufacturing specifications.*

To ensure that government regulations, critical characteristics, and customer requirements were addressed, Kobe engineers developed a matrix to relate these items to control factors of how to achieve them. The matrix also showed the relative importance of each entry, making it possible for the more important items to be identified and prioritized to receive a greater share of available resources.

Benchmarking against best-in-class or competitors was also a function of the QFD matrix, which provided targets for various design features.

QFD pervades the product planning, parts deployment, and process planning phases of a new product. Figure 4-2 shows the major elements of a QFD table.

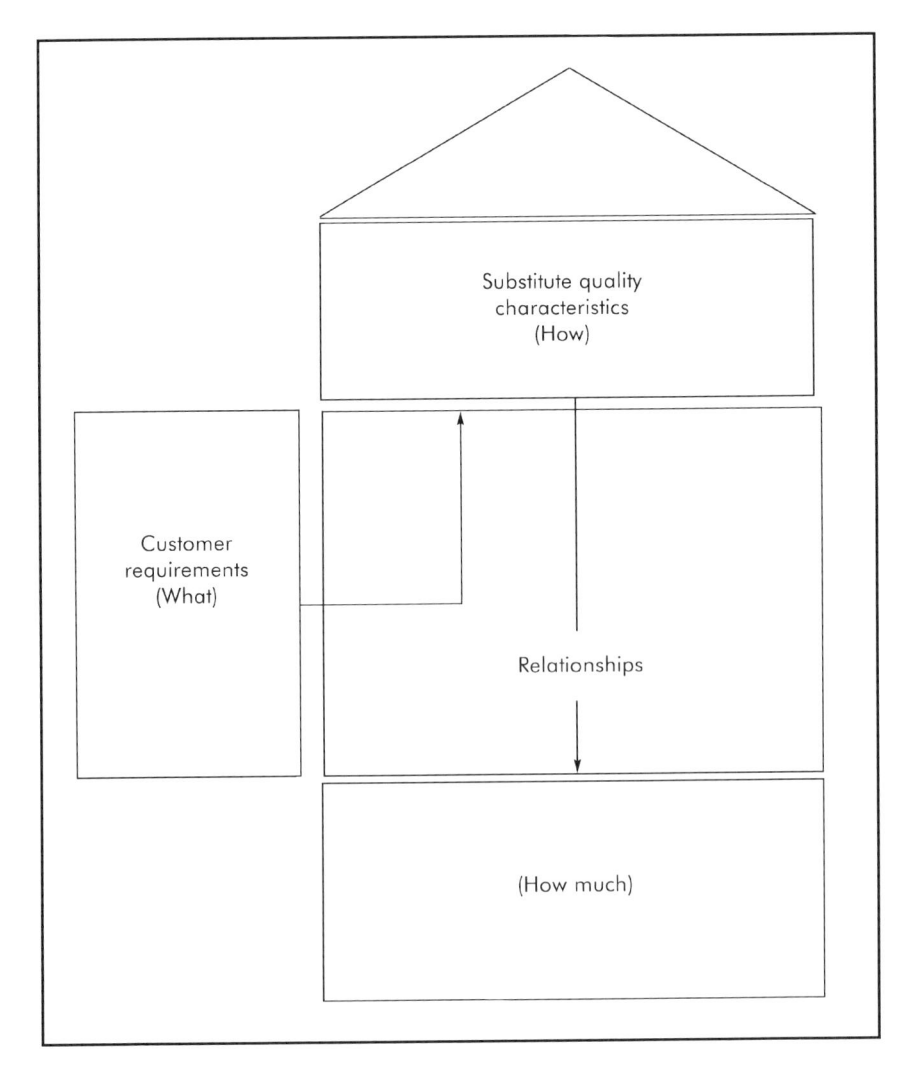

Figure 4-2. Quality function deployment, to be effective, must pervade all major functions of the integrated design/manufacturing process.

The left side of the QFD table shows the customer requirements, or "whats." At the top are substitute quality characteristics, or "hows." These characteristics are intersected with the bottom of the QFD table, or the "how muches." Because the QFD process is iterative and correlates product design to customer requirements, it responds affirmatively to the voice of the customer and provides a measurable competitive advantage.

QFD's success prompted other companies to adopt it in the mid-1970s and the Japanese auto companies were early and successful users of it. In the U.S. automotive industry, it was first applied to a specific problem, rust, a loud complaint of the customer. Since that early rust study, QFD usage has expanded and is now making inroads into many American businesses. QFD, as applied to new product development, is founded on a few basic assumptions:

- Most problems are system related, not people related.

- Market research data traditionally is not focused on the engineering or technical requirements of the customer.

- Real physical and pyschological barriers separating functional areas prevent communication that needs to take place on new product development.

- More time should be spent upstream, early in the process, understanding customer needs and expectations and defining the product in greater detail.

- QFD complements other tools and methodologies such as Taguchi methods, design of experiments (DOE), statistical process control (SPC), and failure mode and effects analysis (FMEA) and is not intended to replace them.

Quality function deployment is a good tool for establishing customer requirements early in the design process. In addition to helping establish minimum design targets, QFD can be used closely with FMEA to identify potential product failure modes, or ask "what could go wrong?" Early in step one, design robustness can be achieved by integrating the voice of the customer, while simultaneously eliminating or significantly reducing failure modes of the design. A robust design is not vulnerable to damage in manufacturing or use.

> *Market research, competitive assessment, and benchmarking have become an integral part of new product development.*

In the automotive industry, QFD is an important tool, since many products must meet minimum design requirements for safety, comfort, fuel economy, and performance. Beyond this, however, most vehicles are sold on paint, trim, and flash, the first-impression attributes of the vehicle that play a major role in customer buying decisions. It wasn't that long ago that features like air bags, ABS braking systems, power windows, power seats, power mirrors, leather seats, air conditioning systems, and styled wheels were only options or were simply unavailable. Today, they have become standard features on many vehicles—a change that was primarily brought about by the voice of the customer. QFD played an integral role in helping to capture these concepts and convert them into designs.

Step Two: Develop Product Assembly, Subsystem, and Component Concepts

After customer requirements have been captured and analyzed, the design process begins. Step two of the process (see Figure 4-1) develops the product assembly system and components concept. It is an iterative process that works extensively with the market research and benchmarking in step one, and the validation of concepts in step three. Step two is where the introduction of computer-aided technologies can enhance the overall concept development and validation through the use of visual engineering.

New concept development involves the evolution of existing designs. Rather than starting with a totally new concept, the design is modified from existing concepts. This is called *continued engineering*. Its premise is that once the concept is defined, predictive engineering can be applied to assess the feasibility of the design.

Continued engineering has been used ever since engineering drawings to describe a concept were introduced. Its purpose is to take an existing design currently used in production and modify it so that it can be used in a new product under development. This assumes (perhaps incorrectly) that the existing part works, it is production-worthy, and the modified design derived from it will work. If it does not work, it is assumed that only minor part changes will need to be made before testing a physical mockup.

Another ill-advised assumption is that companies can make changes and perform minimal testing quickly before release to production. Unfortunately, more changes are always necessary to meet full system objectives, and changes to the physical prototype then must be made and retested. Often, time constraints prevent complete testing for all failure modes and system proveout.

As a result of modifications, final engineering drawings do not accurately represent the design that was actually manufactured, which causes problems and makes the design foundation for the next generation unstable. Final changes to manufacturing processes, independent of the design, are not implemented because they would create an imbalance between "as-designed" and "as-manufactured" products, even if they would improve the product or reduce costs.

CAE technologies have changed the way companies conceptualize, design, verify, and launch new products. Rather than build physical prototypes of the concept and test for fit and functionality, the design concepts can be processed through predictive engineering to examine alternatives quickly and inexpensively.

> *Visual engineering technologies can shrink new product development time by more than 30 percent due to reduction in physical prototypes and testing.*

Moreover, the use of visual engineering and solids modeling techniques allow, for the first time, a common view of a new product concept or design. For example, engineers usually think in terms of geometry, while manufacturing people think in terms of features. Visual engineering and solids modeling bridge the two worlds. With the design concept in a common data base accessible to all with a need to know, much less risk of confusion results. Everyone, including customers, can understand the conceptual design. More importantly, the engineers have their geometry embedded into the realistic design images, while the manufacturing people have a visual representation of the design, from which they can determine the product's manufacturability.

To support the time-based needs of engineers today, faster and more accurate methods of conceptual design and analysis are required. In addition to computer-based tools, a computing infrastructure based upon product information management (PIM) is needed. This computing infrastructure includes a technology bookshelf of previous designs, comparator models, competitive data, design standards, product information, and a process vision that incorporates concurrent engineering.

In developing an initial solid model, the engineer accesses previous designs or comparator designs from the data base. A solid model of the assembly, subsystem, or component is developed using actual engineering geometry, and the volume that the concept can reside in (packaging envelope). This results in a realistic image of the part in its final form.

Concept-level analysis, done on the supercomputer, is typically driven from the solid model using finite element and finite element difference analysis codes to generate the predicted functionality of a design. Future technology evolutions of visual engineering and supercomputer power moving to workstations will allow conceptual designs to be analyzed by the engineer at his or her desktop to simulate operating and experimental bench test conditions.

Step Three: Validate Product Assembly, Subsystem, and Component Concepts

Step three of the process analyzes and then validates, or meets design specifications for the assembly, subsystem, or component design. As shown in Figure 4-1, the design is now ready for development of product and process (manufacturing) requirements. This is where the traditional wall between design engineering and manufacturing must be torn down if robust product development is to be achieved.

Validation of a product assembly, subsystem, or component design takes many forms since it must meet both internal and external validation requirements.

Internal validation takes two basic forms. First, basic product design targets for form, fit, and function must be met. Once this occurs, the design can be iterated through analytical methods to improve its durability, stress, serviceability, etc.

External validation includes regulatory (federal, state, and local) requirements such as environmental compliance, fuel economy, crashworthiness, occupant protection, and safety. In the automotive industry, for example, safety requirements alone make up about 30 percent of all new engineering design requirements.

Continuous price/performance breakthroughs in engineering computing technology have greatly enhanced product design validation.

Today, many companies still use the costly and time-consuming method of constructing physical prototypes to validate product concepts. Physical prototyping and testing consume more than 25 percent of their new product development budgets. Technology is now available to validate and test products using computer-based simulation models which, in effect, are digitized prototypes. Physical prototypes are used to validate the simulations, resulting in substantially fewer prototypes needed to validate the design intent. This reduction in the number of prototypes dramatically shortens the cycle time (and cost) to develop the design concept.

In the 1950s, time was not a business driver in most U.S. manufacturing-based industries. Physical prototypes were constructed and tested over and over until a new product design met initial design targets. Customer requirements were mostly an extension of an engineer's specification, not a driver of the specification. Manufacturing input was not vital, since direct labor was plentiful and cheap. In the auto industry, the government requirements were minor compared with today's complex safety, emissions, and damageability laws, and product liability was an insignificant risk compared with today.

> *The world seems smaller and much faster today, forcing many companies into a race for survival.*

Moreover, global competition was not a concern 40 years ago, since Europe and Japan were mostly rebuilding from the devastation of World War II. Today, the use of high-speed communications, satellites, and computers has virtually shrunk the globe and created a new industrial imperative: to compete with the best products and processes in the world.

Time, however, is of no value if it is not used effectively and efficiently. This means that reducing product development time for the sake of reducing time has no value if the time is not used to improve the quality, cost, or serviceability of the product.

The increasing price/performance breakthroughs in computer-based technologies will help provide engineers with the capability to reduce time, while simultaneously increasing the robustness of the design.

Product validation is a very costly and time-consuming phase of product development. Though technology is available to help reduce the cost and time of product validation, its implementation has been disjointed and focused on individual applications. The key to realizing the greatest potential of the new technology is to integrate it into an overall time- and computer-based *process* for new product development.

This process is founded on the information explosion and the technology breakthrough developed to accommodate it. The information revolution has created a new industrial revolution driven by sophisticated communications technologies. Manufacturers have to process more information today than ever before. However, with its proliferation of local and wide area networks linking workstations and personal computers, that information is readily available to every department, function, supplier, or customer needing it. This information explosion and the technology tools to manage it has, in effect, enabled companies to *integrate* their functions. But though the technology is available, it doesn't automatically follow that organizations will integrate. To make integration happen, new thinking is needed in a much broader context than engineers have been used to in the past.

Going beyond traditional organizational boundaries, the *integrated* approach incorporates the continuous improvement concept into a system, based on the entire value-added chain conducting business as a single unit called an *enterprise*. In the product development domain, it obliges engineers to think on a global level, continue to act on a local level, and throughout be ever aware of the impact their actions and design decisions have on upstream and downstream functions. With the design engineer, manufacturing engineer, analyst, process engineer, and customer engineer involved concurrently throughout the design phase of the product, and all working from a common solids model data base, seamless integration will be achieved. More importantly, though, time to customer and time to profit is shortened considerably.

> *Time-based product development has been enhanced through the price/performance breakthroughs of computer technology.*

The use of solids modeling allows engineers to perform predictive engineering on early design concepts. The use of workstations as the engineer's primary tool is an enabler that has replaced the need for designers and draftspeople to perform part of the function iteratively.

Validation of the product design has been greatly enhanced through the use of computer-based technologies that allow initial analysis, test, and simulation to be performed on the engineer's desktop. This iterative design process results in the creation of a conceptual design that must be tested against target specifications to ensure that it meets or exceeds customer requirements for quality, value, comfort, and timeliness.

Predictive engineering is the use of a mathematical model defined by the design concept to simulate the physical environment and response of the concept in that environment. For simple concepts, closed formulas can be used; however, most simulation is performed using CAE. Figure 4-3 shows that after a design concept is developed, it is transferred from the workstation to a supercomputer through time-sharing.

The supercomputer performs CAE analyses to predict the performance of the design. CAE results are returned to the engineer in a form mathematically rendered to look like a real image. This process requires days to weeks, during which design changes commonly occur, making the engineering predictions no longer valid.

To reduce time in the process and obtain faster results, the envisioned process removes the designer and performs most of the analysis on the engineer's desktop. This is where the "virtual engineer" develops and validates the assembly subsystem or component on the computer. Figure 4-4 shows the current process for performing predictive engineering.

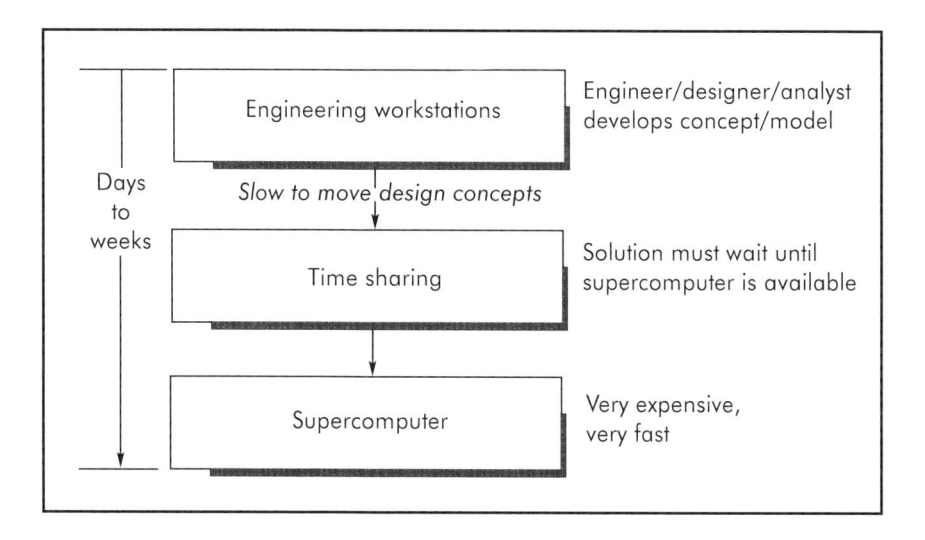

Figure 4-3. Performance of the design is assessed through CAE analysis, a process that can take several weeks.

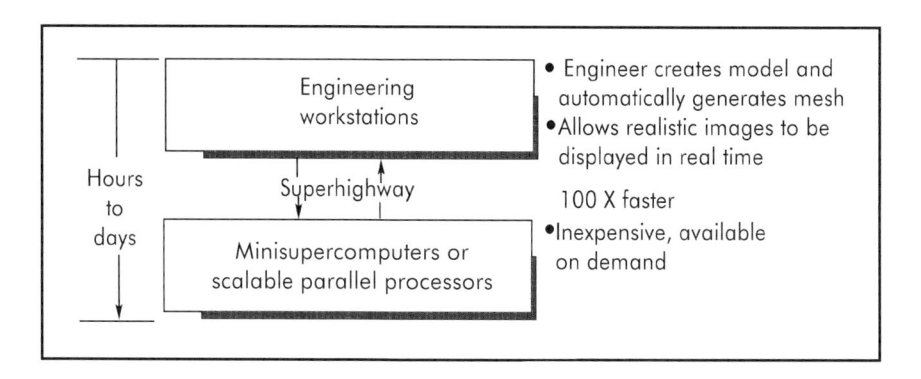

Figure 4-4. With new technology now available, analysis of a product design can be conducted on an engineer's desktop workstation, reducing validation time exponentially.

The new system consists of engineering workstations, high-speed communications, and supercomputers, including the radically-new, massively parallel processor (MPP) systems. This allows the engineer, without waiting, to perform analysis on demand. The system optimizes the entire product development process rather than independently optimizing subprocesses. Control of the product development process returns to the engineer, producing massive gains in quality, cost, and timing.

Validation of the product design is a major milestone event. In the time-driven computer industry, much of the product design validation is done through simulators that predict the performance that results from the integration of multiple integrated circuits. A major product flaw uncovered at this stage of design can usually be captured and resolved easily.

In late 1994, Intel received adverse publicity when it was discovered that their newly-released Pentium chip produced mathematical errors on certain calculations. Intel obviously missed this fault in their validation stage of the Pentium's development process. Intel eventually fixed the problem at a cost reported to be $1 billion, underscoring the importance of product validation and the high cost of error. The Pentium chip had to be not only cost-effective, but also fault-free. Customer perception is critical. Moreover, news travels very fast via media such as the Internet.

> *New product validation must consider that the negative consequences of product flaws may be as tangible as benefits.*

The initial cost of developing a computer-based prototype is less than a comparable physical prototype. Moreover, variations on the computer-based prototype can be reiterated to explore all possible failure modes and to evaluate many design alternatives. This robust design and test capability provides for improved product quality, much faster, at greatly reduced cost.

The increasing capability of solids modeling and visual engineering permits engineers to immediately and visually see their changes. They no longer have to send their designs off to an analyst at the supercomputer to perform simulations and testing. Many applications today provide a very good basic level of design, analysis, test, and simulation as part of the program's functionality.

> *Computer-based prototyping will become a standardized approach to doing business in most product development efforts.*

Once a design has been developed, tested, and validated, it is critical to get early involvement of manufacturing to ensure cost-effective production. Early involvement is critical to developing low-cost designs, since manufacturing cost is often sunk into the design after its release from engineering. Such collaboration is not difficult. Again, the common denominator is the single logical data base and communications standards to facilitate data transmission. A single current version of a design available for manufacturing assessment *during the design phase* will eliminate schedule and cost penalties downstream. In early 1995, Intel reduced the cost of their newest chip by more than 40 percent this way.

Step Four: Develop Product and Process (Manufacturing) Requirements

Step four of the process, as was shown in Figure 4-1, is often called simultaneous engineering, concurrent engineering, or concurrent product and process development.

Early involvement of manufacturing, key suppliers (both production parts and process equipment), and service personnel are critical to the success of a new product design.

Integrating their voice early on in the development process helps to ensure that a robust and cost-effective design is created.

Industrial reorganizations have begun to take place that integrate formerly separate functions of the process in a single location; this is called collocation. Rather than optimizing subsystem performance, the direction has changed to examine the entire process for a product. The American auto industry adopted a *platform-driven* or *-centered* philosophy rather than separate functions, such as body engineering, chassis engineering, and powertrain. The transition to the platform or centrally-controlled concept began in the mid-1980s and continues. Nevertheless, technology, not to be outdone, is changing the work environment again.

The emergence of communication systems as a leading and rapidly changing resource provides an enabler to simultaneous engineering. The ability to transmit pictures in real time allows remote conferencing to be a practical alternative to complete physical collocation. Suppliers and corporate resources located throughout the world become active, real-time participants and members of the team.

Visual engineering is a strong enabler. Through its use as a powerful and realistic communication medium, visual engineering helps to communicate designs and thoughts quickly among the various process stakeholders. Manufacturing can assess the production feasibility on assembly, subsystem, or component features. This quick reaction can pinpoint issues and concerns for the manufacturability of aspects of the design, and changes can be made while the design is still in an electronic state. Production part suppliers can also get a sense of the parts they are being asked to quote on, thus reducing the overall time and cost of procurement. Much of the design's geometry can be reused for free-form fabrication or rapid prototyping of the part for customer evaluation. Process equipment suppliers can provide early input into critical part characteristics that can make the assembly, subsystem, or component much more cost-effective to manufacture while increasing quality. Serviceability engineers can assess design changes for the feasibility of reducing or eliminating costly warranty issues and concerns.

> *Visual engineering helps to tear down the wall between design engineers and manufacturing engineers.*

All new products have a breakeven and time-to-market factor that ultimately determines their success or failure from a financial standpoint. Figure 4-5 shows how Hewlett-Packard Company (HP) views the release of new products.

In HP's industry, product life cycles are measured in months, not years. New products become obsolete within months after introduction due to technology advances. Early design involvement of manufacturing, suppliers, and

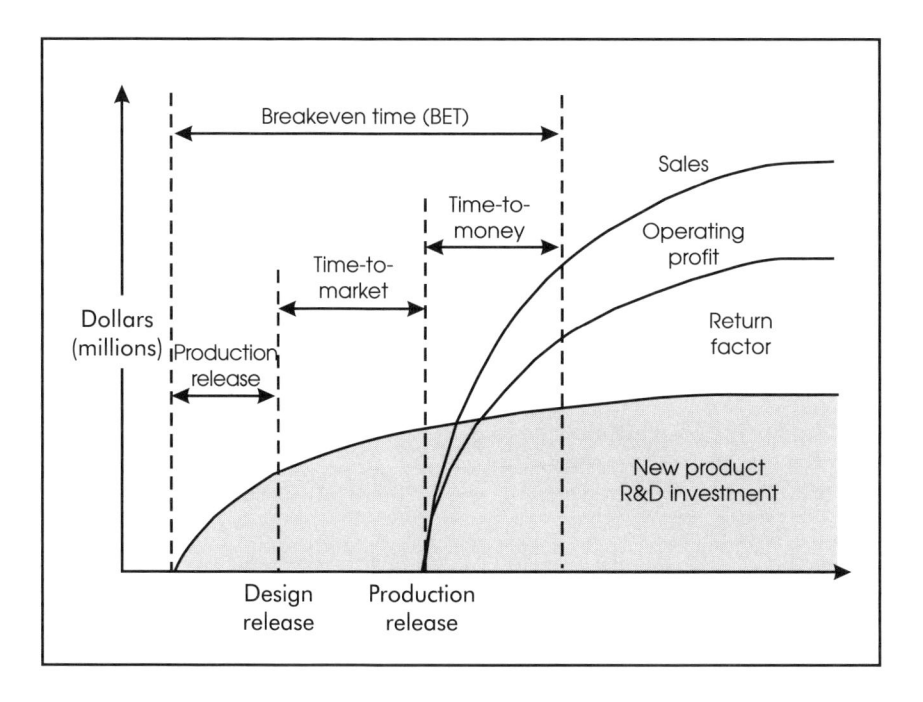

Figure 4-5. This chart depicts how one company plots the product development process in terms of time, cost, and profit. (Courtesy Hewlett-Packard)

field service is necessary if HP is to achieve return factors to support the company's investment in new product research and development.

Figure 4-6 shows the relative growth of engineering change cost as a new product goes through development phases prior to its introduction. Engineering changes, if made early in the process while the design is still in an electronic state, can often cost very little. Changes made later in the process can be 100 to 1000 times more expensive.

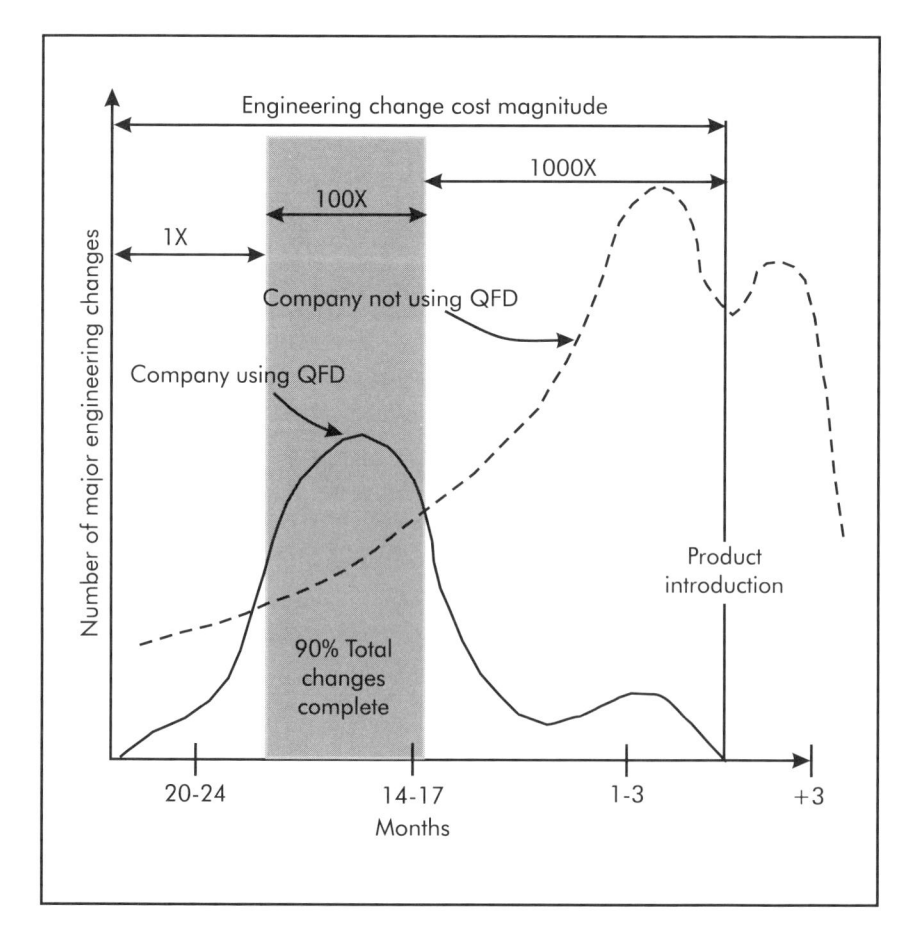

Figure 4-6. The magnitude of cost-per-engineering-change escalates almost geometrically the farther down the development cycle that changes are made.

Step Five: Validate Product and Process Requirements

Most companies have formal procedures for validating product and process requirements. In step five (see Figure 4-1), the product and the process are validated, and the design is prepared for manufacturing.

> *Product design validation is critical to verifying functionality, reliability, and durability requirements of a given design.*

With the advent of simultaneous engineering, or concurrent development, new product development programs are managed as more of a synchronous process. Up-front engineering and product validation provide opportunities to significantly reduce cost, improve quality, and improve timeliness. A good product validation process specifies tests, acceptance criteria, sample sizes, and completion dates for the assembly, subsystem, or component design to be validated. The overall goal of these tests is to verify the functionality, reliability, and durability of a given design.

Design validation testing is normally done with prototype parts that best represent the design intent of future production parts. The prototype is intended to validate the critical and significant design requirements of the assembly, subsystem, or component.

Only well-planned testing that accurately simulates the spectrum of customer usage will ensure that the design is functional, durable, and will perform to the customer's expectations for quality and cost. It is for this reason that product validation should be tied to the QFD matrices.

The process of manufacturing validation for a new product is critical to quality, cost, and timing. The first step in planning for process validation is identifying the critical characteristics of the manufacturing process. The following sources of information are often used to evaluate manufacturing process alternatives.

- Product design failure modes and effects analyses (FMEA), which establish critical characteristics of the product design that must be considered.
- QFD matrices linking critical manufacturing characteristics to critical product design characteristics.
- Process FMEAs, which establish critical characteristics of the various processes used to meet product design intent.
- Process flowcharts, which depict the production process from raw materials through manufacturing and shipment to the customer.
- Computer simulation, which can be performed to highlight production process bottlenecks and schedule scenarios.
- Computer-based, free-form fabrication technologies, which can rapidly generate prototypes for process validation.

Process validation drives the team to investigate the potential of alternative processes as well as their capability under various product scenarios and production volumes and mixes. From these analyses, processes can be selected that optimize the quality and cost objectives of the product. Additionally, action plans to meet these objectives can be developed.

The use of computer-based technologies, such as solids modeling, provide free-form fabrication service bureaus with the geometry necessary to develop production prototypes for analysis and process validation. This can cut weeks and even months out of the overall process. Process validation is normally done with prototype parts and process equipment similar to designs of future production parts and process equipment. The continued advancement of computer technologies will someday allow engineers to virtually "walk" through their factories and assess various production alternatives under various operating scenarios. Today, it must be done with teamwork, shared objectives, a lot of hard work by many, and most importantly, with a passionate leader who balances the needs of all stakeholders.

> *Process validation is critical to verifying what production processes can best meet the cost and quality goals of a given design.*

General Motors and Ford will lose over 50 percent of their existing work force over the next decade to retirement and attrition. One alarming effect is that their knowledge and experience in many aspects of the design and manufacturing process will be lost. But it doesn't have to be. Once the current infrastructure of CAE and CAD/CAM is firmly entrenched, the challenge will be to capture the knowledge of key people, products, and processes so that it can be reused and continually enhanced for use by future generations.

> *Knowledge-based engineering in the year 2010 will be to companies what CAE/CAD/CAM is in the decade of the 1990s.*

Knowledge-based engineering is a discipline that builds upon a computer-based engineering infrastructure, design and manufacturing best practices, and visual engineering. Today, simultaneous engineering requires sheer brute force to get design engineers and manufacturing engineers in the same room. In the future, the voice of the manufacturing engineer will be embedded into design tools that will help guide the development of new designs through parametrics, design cost guides, and a manufacturing- and feature-driven data base of best design practices.

The service and aftermarket industry is also changing as products become even more sophisticated and difficult to diagnose and repair. The computer-based evolution that is dramatically changing the automotive design and manufacturing process is also spreading to the service bay at the local dealership. Service bay diagnostics systems, in the form of computer data bases of product bulletins, service manuals, early warning systems, and increased use of microprocessors in the vehicles, have ushered in an entirely new and very sophisticated form of product service.

> *As leasing of vehicles continues to grow, the trend will be toward free 90-day checkups at the dealer.*

The integration of aftermarket service and support into the design process will provide feedback that has long been ignored by engineers. Most companies traditionally design the product and then throw it over the wall to manufacturing for production, without thought about service.

Service people were never considered part of the new product development team until recently. In fact, manufacturing and service personnel can really make a positive impact on product design by becoming more involved in the critical design activities upstream. Integration will make it happen by affording the capability to plan, design, and execute activities and data management functions up and down the manufacturing chain.

Using concurrent engineering (CE) to illustrate, firms committed to CE have made a big investment in modeling technology, because modeling provides the design-manufacturing-service team a concrete, comprehensive vision of what needs to be done. Many of the decisions formerly in the domain of individual managers are made by a team of engineers, designers, people in manufacturing, and others involved in the life cycle of the product. It is a tool that provides a way of actually writing down a description of what needs to be done in terms that are understood by all members of the team. Modeling also provides a means to demonstrate the need and the effectiveness of proposed solutions, without the cost of implementation, since most of the models that have been developed can be evaluated through simulation. Finally—and importantly—modeling facilitates consensus.

THE WAY OF SUCCESS

- Product development is a series of very structured and highly interdependent steps involving many solutions, without the cost of implementation, since most stakeholders. These stakeholders may be members of the same firm or of external organizations.

- The new product development process is time-compressed, robust, contained on the desktop, and connected globally via the network. It is linked upstream (advanced engineering) and downstream (manufacturing). It drives the rest of the enterprise with real CAE data.

- Converting customer requirements into engineering and manufacturing specifications is critical to the success of the development of any new product. The finished product should be a tangible manifestation of the voice of the customer.

- Companies have had a tendency to design a product and then throw it over the wall to manufacturing, yet to reduce time-to-market and cost-to-produce, product and manufacturing engineers must have the ability to simulate the product's intended environment at the concept level of design, both "in-service" and "being manufactured."

Systems engineering must be used to identify and fundamentally change the existing process to achieve order-of-magnitude improvement.

- The rapid price/performance breakthroughs in computing technology enable companies to further their quest for improved quality and reduced cost.

- The product design and manufacturing processes must be validated early in the product development process.

- External stakeholders such as product part suppliers, process equipment suppliers, and service personnel must be part of the product development team. Yet goals can differ significantly, especially when stakeholders are external to the company. Goals of a supplier firm vary greatly from those of its customer firm. The customer wants to keep the price down, while the supplier wants to keep profits high. What has to take place is negotiation and compromise on short-term benefits in the interest of long-term profit for all stakeholders.

5

2000 and Beyond

The 21st century will be marked by an increasing demand for global products that serve global markets. Fast product development that meets local customer desires for value, quality, and function will be a competitive differentiator for all companies.

PRODUCT DEVELOPMENT PRACTICES OF YESTERDAY

Against a backdrop of rapid, continuous change and unrelenting competition in markets that are now global in scope, manufacturers find themselves challenged as they have never been before. Conventional approaches that have historically worked so well for American manufacturing no longer serve us.

> *A generation ago, most new products were derived from existing products—with little forethought to the customers of the product.*

In the 1970s and 1980s, with the emergence of a budding shop-floor automation industry, our approach to problem-solving was to attack with applications-based, computerized solutions—for the most part, with insufficient experience in the technologies we were implementing. We failed to comprehend the *strategic potential* as *systems capabilities* of the information technologies became available and foreseeable.

A generation ago, engineers solved problems utilizing drafting boards, slide rules, and *Mark's Engineering Handbook*. Knowledge delivery consisted of reference books piled next to the large drawing boards. Most new products were derived from previous designs. Upon completion, a new drawing defined the concept. Until the advent of global competition, this process worked—especially when it was conducted in partnership with key suppliers for engineered products that were manufactured for sale and use in the United States. Production part suppliers were consulted early and often. Key manufacturing equipment suppliers participated early in product design. Engineering and knowledge reuse was not a buzzword, but an actual way of doing business. Purchasing was an active partner serving the engineer and supporting the process.

CAD/CAM changed product development from engineering a concept to stressing packaging and building of prototypes to validate design. Emphasis was placed on designers and their ability to develop alternatives to meet CAD requirements. Product development emphasis shifted from engineering the product for manufacturing to using existing drawings to modify designs that would fit in the design space (referred to as packaging).

> *CAD/CAM ushered in a whole new way to rapidly design, develop, and validate products for manufacturing.*

If the design did not meet the objectives after physical prototypes were built and tested, engineers were brought into the process to analyze the design failure modes and their potential causes and effects. Such failure analysis required specially-trained engineers, called analysts. The analysis process also required long lead times and expensive calculations using costly, high-powered supercomputers. The total process required months or years and was extremely costly. The designer and test engineer drove the process and the product/manufacturing engineers were moved to the later stages of design. Today, computer equipment costs have dropped by orders of magnitude, allowing the process to fundamentally change and return to the basics of product engineering.

THE ENTERPRISE AS A SYSTEM RUNS ON INFORMATION TECHNOLOGY

Computer-based technologies for hardware, software, and high-speed communications form the kernel of process change methodologies that play a major role in the quest for increased engineering productivity.

> *Computer-aided technologies are almost becoming commodities and management needs to leverage their use throughout the business.*

Management must change its focus from considering information systems as fundamental across the entire corporate enterprise, to considering the entire enterprise as a system. Emphasis must change from managing expensive computer systems to developing processes that produce leapfrog quality gains in products manufactured globally. The cost of implementing computer systems must be refocused to the cost of producing quality products. Affordability, although important, is not the key issue. The critical issue is what it will take to stay in business. This will require continuous examination of the business structure as a total system.

NEXT-GENERATION COMPUTER-AIDED TECHNOLOGIES — THEY'RE VISUAL

Figure 5-1 shows that, as processor performance for workstations dramatically increases, the cost of a CPU decreases,

92 *The Virtual Engineer*

CAE Hardware Cost Projections

Assumptions: As processor performance for workstations dramatically increases, the cost of a CPU decreases, allowing additional memory and disk space. The math model can substantially increase in complexity and size, resulting in more accurate analysis for the same cost.

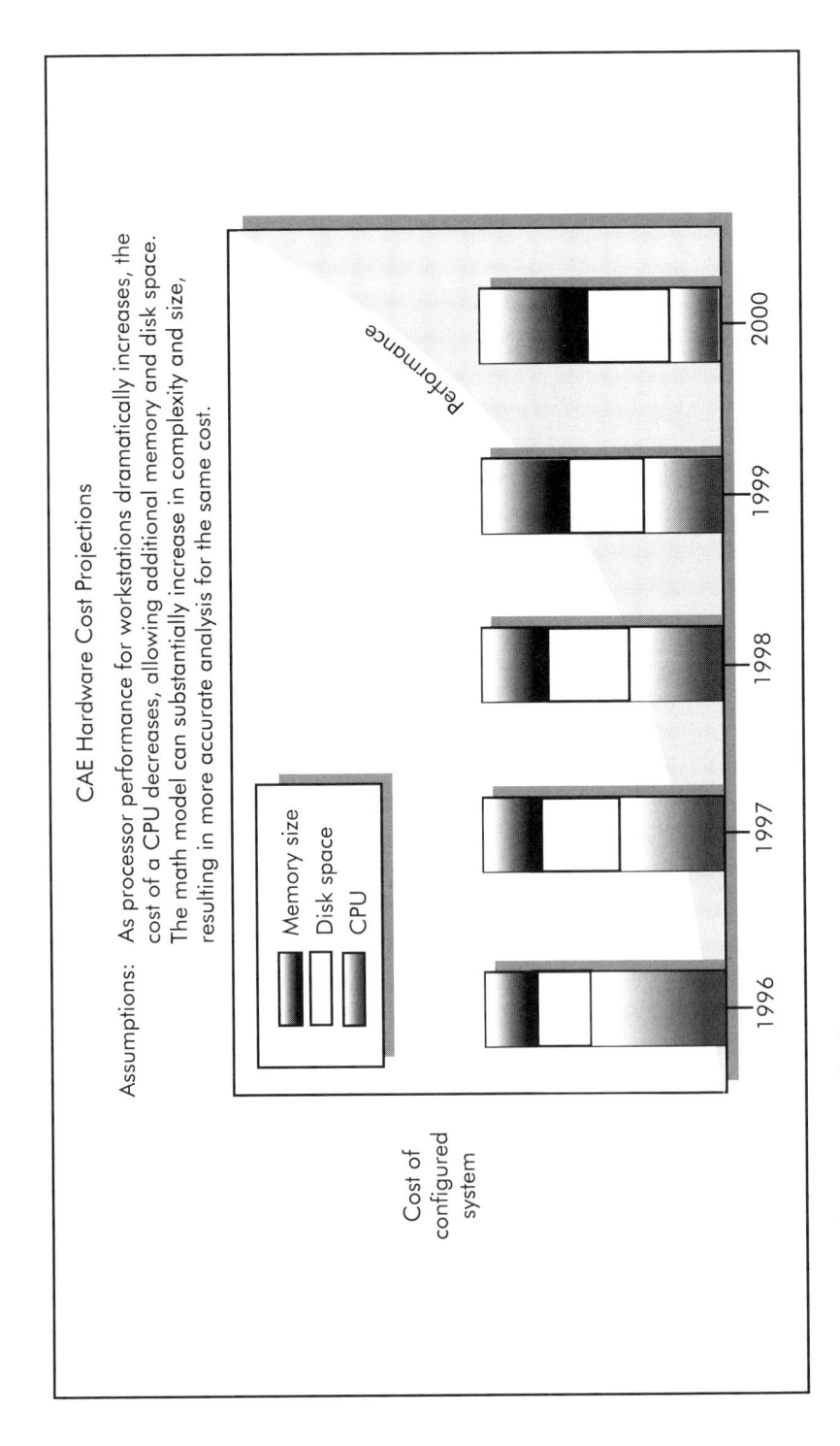

Figure 5-1. Hardware cost projections.

allowing additional memory and disk space at the same cost. The mathematical model used by the engineer can substantially increase in complexity and size, resulting in more accurate analysis. It can be seen that the power being made available to the user is almost unlimited. The next five years will see more major breakthroughs in computer-aided technologies than in the last 50 years. Personal computers will become personal workstations as the continued blurring of microprocessor price/performance ratios virtually make PCs and engineering workstations the same cost. Costs will drop by orders of magnitude. A few years ago, high-end engineering workstations cost about $100,000. Today, systems with as much power or more are under $10,000, and within this decade will be under $1000.

Productivity improvements and new capabilities for both product development and manufacturing engineering will emerge and become the norm. Figure 5-2 indicates these new productivity improvements. The productivity of the product engineer and the manufacturing engineer will double, with less need for specially-trained analysts. Product designers perform analysis and manufacturing simulation with database systems that include analysis tools and previously-designed components.

Engineering Analysis and Testing

Product and manufacturing engineers must be provided the capability at the concept level of design to simulate the product's intended environment, both in service and being manufactured. This ability, called *predictive engineering,* allows simulation to drive the process. Predictive engineering enables the engineer to quickly and accurately analyze design concepts.

Analysis performed in the early 1990s could easily cost $75,000 to $100,000 per evaluation. Additionally, it often could take more than a month to obtain the answers that could only be interpreted by a trained specialist. To perform an analogous study, building a physical prototype and testing it, would require about 4 to 6 months to construct

94 *The Virtual Engineer*

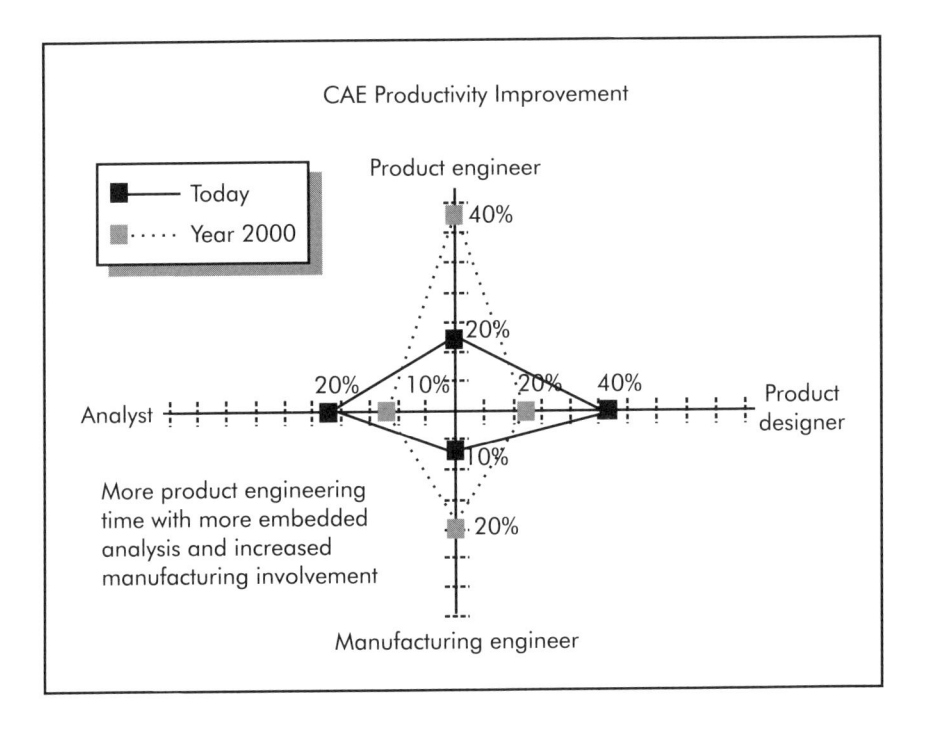

Figure 5-2. Engineering productivity improvement.

and about $500,000 for the test. The advantage of building the physical prototype was that film analysis and the table of results could be interpreted by an ordinary engineer.

> *Management has traditionally been more comfortable with physical test results than analytical models.*

In the past, management usually preferred physical tests because they were comfortable with results they could see. Using the systems engineering approach, an engineer now spends 2 to 3 days preparing the design model, simulating the model under load conditions, and reviewing the results. The cost of this simulation, with its visual display, is less than $2000. These dramatic changes in time and cost result in analytical prediction replacing the traditional build/

test, and represent a fundamental change in how a concept is designed. This breakthrough returns to a tradition of performing engineering and design together. It enables companies to return to the basics of people, process, and technology and apply them systematically to continuously improve engineered products.

Visual engineering, the representation of design concepts as realistic images, coupled with the solids model (3-D) representation revolutionizes the design and manufacturing process. *Solids modeling* is the exact mathematical representation of design concepts in a 3-D representation. It is based on 2000-year-old Euclidean geometry—hardly a new technology. The number of calculations between the human eye and brain is 10 billion per second. Visual engineering is the enabler of this powerful capability.

> *Engineers prefer the interpretation of 3-D and solids models once they have been exposed to the process.*

Next-generation Engineering Workstations

The enabler for 3-D and solids models is the low cost of workstations and great power of the latest software. They allow the user to perform complex calculations in microseconds, a blink-of-the-eye, and represent the concept in a realistic surface form (visual engineering) rather than lines (see Figure 5-3). Once the solids representation is used, few engineers ever want to return to the 2-D Euclidean representation.

The solids model representation also enables dramatic new changes:

- Automated meshing that allows for the concept to be prepared for simulation without human intervention.

- Simulation models that yield accurate answers repetitively.

Figure 5-3. Solids representation—design concept.

- Direct manufacturing of the concept design using rapid prototyping and paperless machining.
- 3-D printers in place of traditional 2-D printers, giving users immediate tactile sense of the design.

The engineer of the future will have the equivalent of a supercomputer and product analysis laboratory (Figure 5-4) on his or her virtual desktop. The network with standards, such as ACIS/Parasolids/STEP for seamless connection, will give the engineer access to concept design, packaging constraints, predictive engineering, and rapid prototyping from the desktop workstation, using a single solids model. Virtual reality and holography will be used increasingly and become an enabler for the development team of engineering, marketing, manufacturing, and suppliers in real time and the real world.

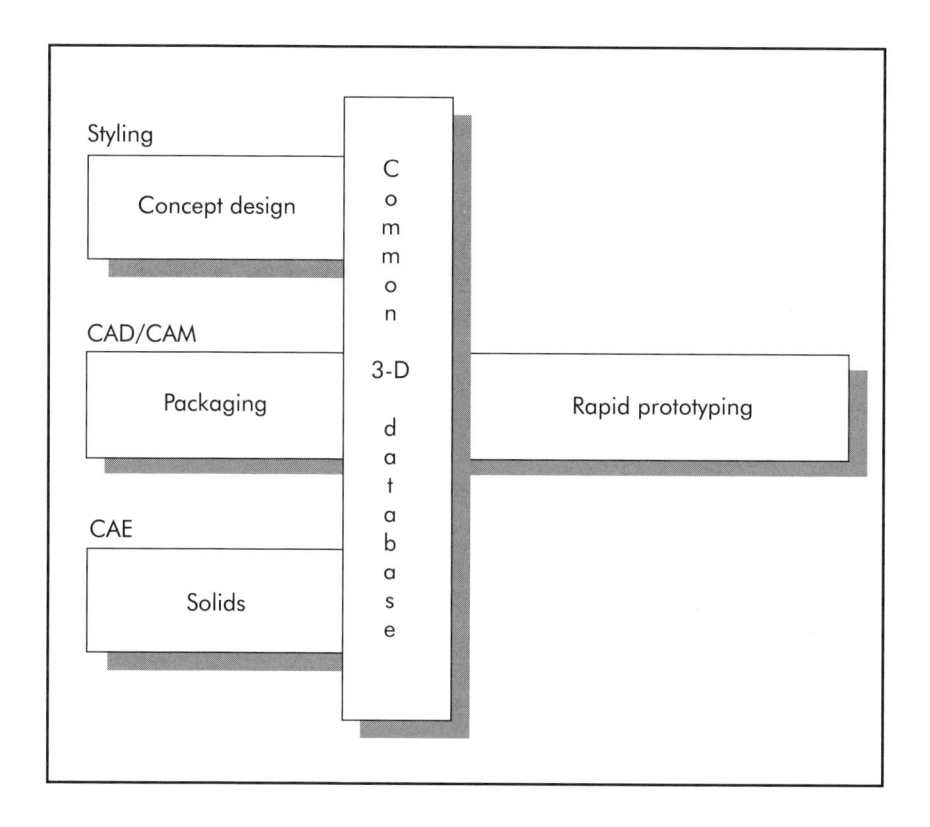

```
                    Styling
                              ┌─────────────┐  C
                              │Concept design│  o
                              │             │  m
                              └─────────────┘  m
                                               o
                                               n
                    CAD/CAM
                              ┌─────────────┐  3-D        ┌──────────────────┐
                              │  Packaging  │             │ Rapid prototyping│
                              │             │  d          │                  │
                              └─────────────┘  a          └──────────────────┘
                                               t
                                               a
                    CAE                        b
                              ┌─────────────┐  a
                              │   Solids    │  s
                              │             │  e
                              └─────────────┘
```

Figure 5-4. The virtual desktop system.

> *Next-generation engineers will have virtual product development laboratories on their desktops.*

The requirements for performing engineering studies will shift from supercomputers to inexpensive desktop systems by the year 2000. The migration from supercomputers will be to massively parallel processor (MPP) systems as compute servers, as well as MPPs moving directly to the desktop workstation. MPP technology will be implemented on the desktop in less than 5 years and 95 percent of the problems that require supercomputers and MPPs will be solvable on the engineer's desk. Large processing capabilities will be developed on the workstation, and the network will connect the user to the necessary resources. Figure 5-5

98 The Virtual Engineer

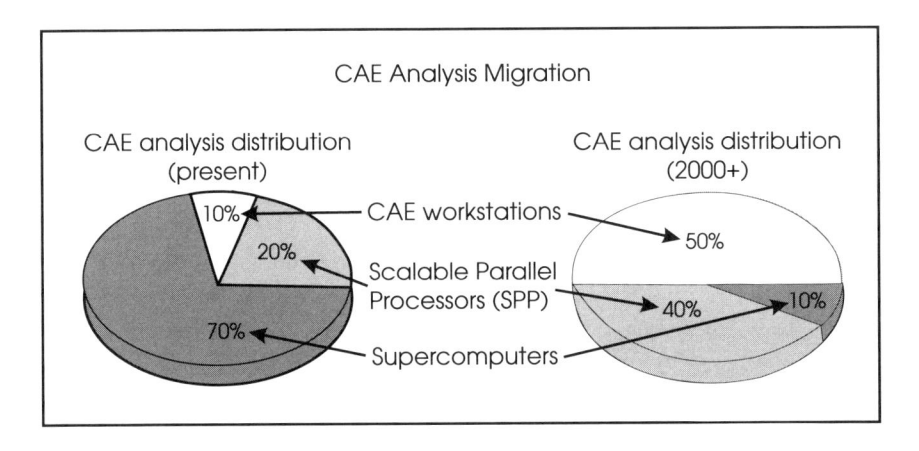

Figure 5-5. Predictive engineering migration to virtual desktop.

describes this migration into the next decade. The need for the supercomputer will dwindle, as CAE workstations become ever more powerful and cost-effective, and SPPs and network technology are increasingly utilized.

Communication Systems

Although PCs/workstations and supercomputers are undergoing unprecedented changes, the greatest change is occurring in communication systems—the networks and related technology. The heart of the information age is the network, supplying information to the necessary arteries quickly and almost in real time.

The critical problem today is transmission speed. Relatively slow now, it will dramatically improve over the next decade to the point where users can transmit color and solid object pictures in real time (referred to as pixels), as well as the parametric definitions that were used to construct the design or 3-D image.

Systems will be Knowledge-based

The ability to create designs and quickly simulate many design changes allows the engineer to develop a new arsenal of knowledge. This will be key to enabling the most dramatic changes to occur over the next 50 years.

Knowledge-based systems have been discussed since the mid-1980s and are based upon engineers' experience. This experience is quite limited at best and does not address optimization for quality, cost, or time. However, knowledge-based engineering represents a set of rules that allows an ordinary engineer to obtain tremendous knowledge of the product concept never possible before. It allows the engineer to structure data in an understandable format, allowing better and shared communication of information.

> *Knowledge-based engineering, combined with solids modeling and analytical prototyping will be the CAE process enabler for engineers. The next evolution will be wisdom-based engineering.*

As experience is expanded to include external inputs from marketing, customers, suppliers, product planning, and corporate direction, wisdom in the form of a greatly enhanced knowledge base becomes the enabler that will give companies improved competitive advantage in the 21st century.

CHALLENGES AND OPPORTUNITIES

Because of its innate unpredictability, the future is both frightening and exciting. The continuous quest for high-quality products, lower manufacturing costs, and faster time-to-market will force many companies to change current policies, processes, and practices for product development, or face the competitive consequences. On the plus side, continued price/performance breakthroughs in computer-based technologies will allow many companies to leapfrog their competition in the product development arena.

> *Companies that manufacture engineered products have no choice but to change the way they develop them.*

It cannot be overstressed that companies must fundamentally change the way they develop, design, and manufacture products. With the ramifications of the global market now reality, companies can manufacture or outsource their manufacturing anywhere to meet localized needs. As high-speed communications and information superhighways shrink the globe, world-class competition is only a mouse click away.

The industrial age in this country lasted over 150 years and was unceremoniously buried with the advent of the microprocessor in the 1950s. The information age has witnessed continual price/performance breakthroughs in the microprocessor, coupled with new graphic-based software and high-speed data communication.

After the Information Age Comes the Virtual Age

As year 2000 approaches, we are entering a third age—the virtual age—marked by the increasing use of microprocessors in everything we do. The computer has, for better or worse, become a part of the culture, attached to both the person and the job. Our daily lives have increasingly become dependent upon the microprocessor for our standard of living, and this is not likely to change as we enter the 21st century.

The doubling of information processing capability every year has created a path of no return. The increasingly affordable microprocessor, although still controlled by several chip-making keiretsus, is now considered a commodity. The microprocessor is inextricably wired to our daily lives and is showing up in everything from video cameras to washing machines. Our lives have been permanently changed and our standard of living has never been better.

> *Creation of the microprocessor is to the information and virtual age what interchangeable parts and mass production were to the industrial age.*

Neural Net Becomes a Reality

The microprocessor has placed increased emphasis on greater use of an extremely critical resource—the human mind. The U.S. educational system has failed to keep pace with technological changes. New focus must be placed on preparing people for a microprocessor-driven society.

The human brain has over fourteen billion neurons working simultaneously to organize information for our mind. These neurons act on input from external sources—our senses. The neurons are connected to form a neural network to assimilate and organize the data into a format that the intellect can recognize.

External research is underway in the U.S. and Japan to explore and develop computer systems that simulate the brain. The emerging fields are called neural networks and artificial intelligence, both of which have begun to reach the industrial sector in small process-related applications. Initial applications have taken much longer to implement and needed more computer resources than anticipated. As new architectures such as scalable parallel processors (SPPs) replace the traditional systems of today, neural networks and artificial intelligence applications will improve, expand, and grow over time.

> *The integration of artificial intelligence disciplines such as knowledge-based engineering with desktop supercomputers brings visual engineering into being.*

In the next century, neural networks will be a critical resource to many processes, particularly product development and predictive engineering. The computer architecture enabling the simulation of the brain will take the form of software and knowledge-based systems running on computer workstations and MPPs.

> *Even with continued breakthroughs in computer-based technologies, the human mind is still a mostly underutilized supercomputer.*

Neural networks are developed by knowledgeable engineers who establish the rule sets for the query process to select from alternative actions based upon external measurements. From this experience, best practices can be developed to simulate the best decisions.

Artificial intelligence will support an emerging discipline called knowledge-based engineering. Knowledge-based engineering blends the knowledge and experience of design engineers, manufacturing engineers, and design best practices into a product development process that is more robust, faster, and less costly than traditional processes.

The Cost of an Engineer

As we approach the year 2000, we see the equivalent of a single supercomputer on the desk of a highly trained engineer. An engineer today costs more than $400,000 annually, with a lifetime cost approaching $12 million. This is a significant investment, especially since we use less than 10 percent of the engineer's overall capability today.

Engineering productivity must be a focus of the product development process. Metrics, in the form of quality, cost, and time-to-market must also include engineering cost per unit as a measure of real product development performance.

The cost of a fully-equipped engineer in the year 2000 and beyond will approach $500,000 annually. This cost will force many companies to reassess their core values and competencies relative to product development. Product development is a last bastion of competitiveness that will prompt companies to consider their options.

Outsourcing

One alternative finding acceptance with many companies, is that of outsourcing product development and manufacturing and focusing more on becoming global marketing and money managers. The technology infrastructure is currently being put in place to make this a reality.

Electronic Data Systems (EDS) became a major force in the information systems technology business by offering companies an improved information service at an annualized cost that was 15 to 20 percent less than the company could provide internally.

> *Core competencies will come under attack as more and more companies outsource their product development and manufacturing.*

More companies are outsourcing manufacturing, particularly in search of lower-cost or better locations than they could provide internally. Consider United Parcel Service (UPS) and Federal Express (FedEx) which are multi-billion dollar companies today because they can deliver packages and overnight mail to anywhere in the world better and cheaper than anyone, including the U.S. Postal Service.

The logical extension of this core competency assessment and outsourcing process is that product development will begin to be outsourced by major companies. A case in point: major automotive companies were pioneers in developing CAD to replace drafting processes in the late 60s to mid-70s. Drafting was outsourced, spurring the development of the CAD/CAM companies so prevalent today. Today, CAE functions such as solids modeling, finite element and fluids analysis, rapid prototyping, and general engineering support to product development is being outsourced to many contract agency firms. These will evolve into whatever replaces today's CAD/CAM industry.

> *Time-based product development will increasingly look to outside service firms to provide critical resources in computer-based technologies.*

Companies find it hard to change their culture fast enough to meet the time-based needs of the marketplace. Many workers still cling to the notion that they have a right to a job, and this right extends for a lifetime of gainful employment. Unfortunately, those days are gone, and companies will increasingly look to outsourcing jobs, including product development, in their quest for faster and more cost-effective development.

NEXT GENERATION PRODUCT DEVELOPMENT

The year 2000 will witness a virtual product development team that is a homogenous blend of project leaders, design and manufacturing engineers, contractors, production part suppliers, and process equipment suppliers. Traditional management jobs will disappear as managers take on new roles as project leaders for a predetermined period of time, for example, 6 months to 2 years. In addition to managing, they will also participate actively as team members and often will not be the lead in the process.

Resources for these various projects will come from both internal and external sources. More and more of the key product development functions, however, will be increasingly outsourced to minimize staff and capital costs for new product development.

> *Traditional product development teams will look increasingly to contractors, production part suppliers, and process equipment suppliers.*

Capturing the voice of the customer was a focus in the 1980s and 1990s. Customers played a major role in Boeing's new 777 airplane product development. These customers (in the form of airlines, pilots, flight attendants, service personnel, etc.) provided major input into the development of new systems and support procedures for this advanced-technology aircraft.

Designing to the voice of the customer will be standard operating practice in the 21st century. State-of-the-art desktop technologies and visual engineering will enable the product development team to evaluate, build, and test a much wider variety of design concepts in the computer. Knowledge-based engineering will provide input on manufacturing process best practices, facility and tooling data, labor estimating data bases, packaging, material handling, and freight information. Visual engineering, in the form of solids modeling, 3-D printers, and free-form fabrication, will be used as a standard medium of communication between all product development team members. Predictive engineering, as opposed to testing of physical prototypes, will provide a major shift in product design validation, as well as increased quality and reduced time and cost in the process. Testing will still be required to ensure all physical variables are properly simulated. Since the design team is fallible, testing also will be used as the final step before release.

> *Knowledge-based engineering will support the engineer with expert guidance on design and manufacturing best practices.*

Virtual reality will drive the process. Team members will design concepts in *real time* in a *real world* simulation of its intended customer use environment. Dramatic increases in productivity, undreamed of in the 1980s, will begin to replace the traditional processes of new product development.

Centralized and time-shared supercomputers will migrate to desktop workstations, compute servers, and scalable parallel processors (SPPs). Linked by a superhighway network in the form of a worldwide web, these computers will demonstrate a capability to move information at the speed of light (186,000 miles per second). This computing infrastructure will provide the product development team with all of the CAE resources they need to locally control and manage the entire process through elements known as the *virtual desktop*.

Today, SPPs are helping companies stay current with computing technology while increasing their productivity and asset usefulness. Many companies have experienced the death spiral of having to swap out high performance CAE compute servers every year or so to keep up with increasing engineering computing requirements.

> *The virtual desktop of the product development team in the year 2000 and beyond will consist of workstations and SPPs.*

SPPs have a computer architecture that permits an application to begin with a relatively small number of processors (fewer than eight) and scale up to hundreds and thousands of processors. SPPs of the future will enable engineers to develop design concepts in minutes and hours rather than days and weeks.

Scalable parallel processors speed up the development of complex CAE models and give the design engineer increased flexibility to reconfigure a CAE resource system quickly and easily. An analysis that would normally take weeks to complete with traditional compute servers can be completed overnight by reassigning the system's processors to run a specific problem. In addition, the virtual desktop of workstations and SPPs increases the available processing power capacity so users can run large, complex jobs simultaneously on demand.

Computer workstation technology will continue to improve in price and performance. Advances in chip technology and graphic displays will allow more changes over the next 5 years than have occurred over the past 50.

> *Computer workstations will provide a complete suite of high-performance computing and networking capabilities on the engineer's desktop.*

The next generation PC or workstation will be very small and very lightweight, with a high-resolution, flat panel display. The desktop system will evolve into a portable or laptop system. As flat panel displays become more cost-effective, computer manufacturers will move away from the traditional desktop PC to the lap-top.

As chip and display technology advances continue, it is reasonable to project that the virtual desktop system could be configured into a system the size and weight of current portable cellular phones. True simultaneous engineering could take place anywhere in the world at any time of the day in real time.

> *People, and their ability to accept and embrace change, will ultimately determine the success or failure of companies in the future.*

SYSTEMS ENGINEERING—INTEGRATOR OF PROCESS, PEOPLE, AND TECHNOLOGY

Many companies have good people, excellent product and process technologies, and certainly an external push from competition to get better. The key is to step back and take a systems engineering approach to the integration of the people, product, process, and technology in the business.

Despite all of the advances in technology, it appears that people are working harder and achieving less. Have we reached the point of information overload? Do we spend more time looking for information than using it intelligently? Is too much attention placed on benchmarking the competition instead of beating it? Are we focusing on individual solutions when we should be looking at systems strategy?

> *People spend more time today looking for information than using it effectively.*

Systems engineering is critical to competitive success. Rather than looking at each activity as something to be improved independently, the emphasis of the information age should be on examining and optimizing the process as a total, single entity, or system. The trade-off of packaging, performance, manufacturability, and total systems life cycle costs must be made at the system level, not at the subsystem or component level where most companies operate today. Fortunately, manufacturers, particularly in the automotive industry, are discovering systems engineering.

The coming decade will be one of the most exciting and demanding times for manufacturing since the introduction of the Industrial Age more than a century ago. As we enter the next millennium, the Information Age will be well entrenched. Enabling technologies such as solids modeling, parametric design, analytical prototyping, free-form fabrication, 3-D printing, and knowledge-based engineering will provide engineers with the capability to effectively respond to the voice of the customer. Product development will take on a much broader character, involving the concurrent activities of design, manufacturing, and process engineering working in concert with suppliers, internal and external customers, and manufacturing partners up and down the value-added chain of production.

> *Design will increasingly become an art form, with engineers able to see their creations in minutes or hours after concepts are developed.*

Advanced technologies, integrated with more robust systems engineering training methodologies, will help companies achieve their product development goals. Systems engineering will be the guiding discipline to accelerate and allow industry to transcend to the next manufacturing plateau. But to achieve that potential, manufacturers must be prepared to both accommodate and implement massive change. The success of visual engineering, systems strategy, and enterprise integration are founded on the fluid interaction of people, process, technology, and organization, functioning as a single, cohesive unit of production.

WHAT TO LOOK FOR

- The future, although increasingly demanding in the global marketplace, will be an exciting period of change for product development.

- Computer-based technologies—hardware, software, and high-speed communications—will play a major role in achieving increased engineering productivity.

- Companies will increasingly look to visual engineering technologies, such as solids modeling, as a primary communication method for design concepts.

- Virtual reality will increasingly be used to enable product development teams to function as a cohesive unit in a "real-time/real-world" framework.

- Microprocessor technology will continue to outpace our ability to deploy it to meet the needs of the engineer.

- Product development engineers in the year 2000 will have the equivalent of a supercomputer and product analysis laboratory sitting on their desktop.

6

Organization Overhaul

INTEGRATE, INNOVATE, OR EVAPORATE

Integration is the theme that underpins the success potential of new product development in the 21st century. Much has been written and championed about leapfrog advances in technology, the advantages to be gained by integrating them, open systems and seamless communication among electronic devices, and designing products, processes, and service around a common data base. What's not so evident in current management thinking and practice are the radical organizational and cultural changes that are critical to making successful technological integration a reality. Linking technology innovations is only part of the integration equation. Organization is only part, as is the people component. The competitive advantage of integration lies in a context much broader than hardware, software, and process design.

A successful change innovator was Deming who assisted the Japanese industrial culture in changing during the 1960s and 1970s. His principles and processes for examining business strategies are rooted in systems engineering.

Systems engineering optimizes the system rather than its subsystems. Consider the following argument: if we have a system today that we feel is optimized and break it into its primary subcomponents and then try to optimize each subcomponent independently, when the system is reassembled, it probably will not function, and if it is working, it will not be optimized. The corollary is that if you have an optimized system, then your subsystems cannot be optimized individually. This is a major paradigm shift in thinking by senior management and plays havoc with human resource departments in developing performance reviews.

Imagine the president or CEO telling the VP of Product Development that he or she must add time (and cost) to realize a savings to the total process. This is exactly what must happen. Adding time at the concept level to understand its relationship to customer wants, manufacturability, etc. results in fewer prototypes and more efficient manufacturing. The total cost is reduced by increasing a subsystem cost and decreasing another subsystem cost. Systems engineering represents the enabler to obtain leapfrog improvements in the process.

The key is to examine existing subprocesses and understand their individual contributions to the total system process. This requires a mind-set change that examines trade-offs within the subsystems that improve the total system objectives. It requires an understanding of the current process, its key elements, the interrelationships between the subprocesses, and what can be measured at the subsystem level, as well as the system level.

Once the existing process is understood, then changes at the subsystem level can be made and the total system change can be determined. This technique requires that the total system and the interrelationships of each subsystem are understood up front. This task may seem formidable, as manufacturing today is a wide-ranging network of customers, channels, suppliers of components, providers of process technology, contractors, etc. The strategic capabilities of the information technologies available to us must be exploited to develop a systems approach to production.

Conventional approaches that have historically worked so well for us no longer serve us. In the 1970s and 1980s, with the emergence of a budding shop-floor automation industry, our approach to problem-solving was to attack with applications-based, computerized solutions—for the most part, with too little experience in the technologies we were implementing. We were substituting computers for manual operations on a task-by-task basis, not comprehending the strategic potential and big-picture capabilities of the available information technologies. We were concentrating on the trees when it was the forest that was dying.

The results, as we became painfully aware, were hundreds of individual applications—islands of automation that did not communicate with each other, each with its own data, language, and terminology, all producing less-than-expected benefits at considerable cost. A loosely-connected value chain still faced growing competitive pressures to increase overall responsiveness to customers and lower the total cost. Introducing new technology did not by itself really improve results. To use anything new, one must go through the process of *learning* how to use it, and that may mean *unlearning* what one already knows. A classic case in point is how, where once there were vendors and customer companies, there are now partners—a chain of supply and production seen (and evaluated) by the end user or customer as a single entity. The customer can be internal or external. The network can be totally in-house or partially outside the company, it makes no difference. A fluid production network is the key to competitive advantage, and integration, flexibility, and innovation make it work.

THE PAST IS PROLOGUE

Because of a narrow focus in the past, and a lack of knowledge of the potential of automation, strategic planning for the Information Age wasn't considered. Executives were in denial. Only recently has the manufacturing community begun to grasp the overwhelming possibilities of a *systems* approach to production and to fathom its implications.

The manufacturing enterprise today comprises a chain of producers and users who have intricate operational dependencies on one another, both internal and external to individual companies. Huge amounts of information are generated by each and available to each. Unfortunately, these producers and users are not tightly linked (if linked at all) to each others' business operations, and much valuable information goes unused or is used ineffectively. Indeed, in many cases, it is not clear that any one member of the enterprise causes a problem or is responsible for driving a solution. For a major manufacturing company, this creates monumental problems. But out of these problems have come great opportunities.

In any given industry in the U.S., you can go through the reams of management literature produced over the last 40 years and come up with a consistent 3- to 4-year cycle of activity bursts surrounding productivity and cycle-time-cutting issues. Each time these bursts occur, new items, new ideas, and new approaches to improving productivity emerge and capture managerial attention, at least for awhile. Excellent programs have come out of these efforts—statistical process control, quality function deployment, total quality management, design for manufacture/design for assembly, Just-in-Time—to name just a few of the most noteworthy—all promising in concept, effective in practice, and broadly espoused. Collectively, they are making a mark on industrial productivity, and they bode well for the future. To be sure, expansion of such concepts and creation of new ones should be on the agendas of all of corporate America if we, as a manufacturing power, are to survive and thrive in the 21st century. The problem is, they are for the most part implemented as discrete programs to solve localized problems, and not integrated into a global plan as parts of a strategic corporate thrust involving all stakeholders. A product development function priding itself on successfully implementing these programs, for instance, is still in reality only a link in the productivity chain. Functioning without interest in or knowledge of how designs impact operations up and down the manufacturing chain does little to reach the overall goals of reducing time and responding satisfactorily to customer demands.

The links to join the chain exist. The evolution of information technology and development of the tools to exchange that information have provided the capability to expedite the burgeoning information exchange needs. In fact, information communication technology will drive the new industrial revolution. Manufacturers have more information at their disposal today than ever before. With the proliferation of local and wide area networks linking workstations and personal computers, that information can be readily available to every department, function, supplier, or customer having to act upon it. This information explosion and the emergence of the tools to manage it has, in effect, enabled the *enterprise* to *integrate*.

ENTERPRISE THINKING

Though the tools of integration have been available for some time, it doesn't necessarily follow that integration will happen automatically. The two other elements of the integration triad—organization and people—have to be in place and functioning. Simpler organizational structures will have to be created. Flat, fluid, flexible organizations with far fewer layers of control have become the operational model of cutting-edge companies preparing for 21st century success. These structures have at their core a strategic commitment to continuous change and making change an agent of opportunity. But one of the most difficult changes manufacturing executives have to make is in their thinking.

While research labs continue to provide advances in *technological* systems, changes in the *social* system of the workplace receive little attention. Humans tend to be uncomfortable with change. In the case of advanced technology, such as solids modeling and powerful desktop workstations, there are definite disadvantages to its introduction. To use something new, and go through the process of learning how to use it, may mean unlearning what one already knows.

A good case in point is the introduction of robots to the shop floor. Shop-floor workers—especially organized labor—*favored* the introduction of such devices and recognized them as essential to promoting economic progress through increased productivity. Interestingly, it was middle managers who opposed robots most stridently. American workplace culture equates the manager's status with the breadth of control and the number of people reporting to him or her. With a department of robots and fewer human workers, the managers felt diminished authority and importance. The robot's fit just wasn't there. In fact, there was a tendency for managers and other executives to create a layer of human beings between themselves and the robots to preserve their status. For decades, industry in the U.S. has functioned well with such a highly formalized, multilayered structure. The long tradition of fierce independence and rugged individualism made it difficult for some to work as team players. Yielding power, control, and authority for the sake of a better competitive position was a foreign notion.

The new model is much less structured. It eliminates many of the layers bottom to top, and is flat, fluid, and flexible. Roles within the organization are less strictly defined, providing the flexibility to respond quickly to the voice of the customer. Cross-functional interaction is assumed, enabling decision-making authority to be pushed to the operational level where empowerment can be exercised in line with the goal of total customer satisfaction.

TECHNOLOGY, ORGANIZATIONS, AND PEOPLE

In a broad sense, integration can be defined as connecting people, organization, and technology for greater strategic impact and productivity throughout all the components of an enterprise.

The manufacturing enterprise consists of a number of organizations, each contributing a unique, distinctive competency to the global system. They all share responsibility for cost, delivery, and quality, and they each participate in solving problems where they have a comparative advantage in the system.

In such an enterprise, decision-making is vested in local cross-functional, organizational teams who use the most current information to manage production and make on-the-spot improvements. No longer is information about manufacturing activities gathered after the fact into historical reports for senior managers. Translating customer needs into a final product requires efficient information-sharing across the chain of producers and users, whether they are in-house or external. Each operational team controls a local piece of the enterprise and has an impact on the global system. The goal is to optimize the global enterprise, which is only possible when the team has enough timely information to not only manage its own business, but also manage the impact its operations have on other teams and on the total enterprise. Manufacturing today extends far beyond individual department walls, even beyond individual factory walls. In a globally competitive environment where one lost opportunity can sound the death knell for an entire company, time becomes the overriding determinant of whether that company thrives, survives, or dies; integration provides the speed to generate the competitive edge.

Integration builds on relationships, and expands them to provide system-wide efficiencies through certain fundamental elements.

- It supports effective industrial networking. Integration should leverage the strengths and capabilities of a corporation to its suppliers and its customers. This links the value-added chain into a single functioning unit of activity, building effective partnerships with shared strategic goals, work teams, and a mutually sustainable competitive advantage. Integration emphasizes the *value-added* activities of the interacting organizations more than their separate functions, replacing ad-hoc personal relationships among those organizations with business activity teams across the enterprise. It's no longer who you know that links the enterprise, its part of the team concept.

- Second, integration recognizes that an enterprise is a complex system and focuses on those interdependent activities that go outside the boundaries of individual organizations. The objective is to optimize the productivity of the entire system by optimizing the activities of all the individual units in relation to the entire system. It is, in reality, an extension of the notion of continuous improvement across the global organization and its networks of activities. Each unit must function to continuously improve the enterprise; one superstar in a chain of mediocre performers will not carry it.

- Third, integration provides a single data resource for the control and coordination of activities across the enterprise—a single, current copy of information. Design, for instance, is a process of managing information and decisions around that information. Using concurrent engineering (CE) as an example, a necessary component for supporting CE across an enterprise is the availability of a single, current version of product information that is accessible to all who deal with it.

- Finally, integration provides a foundation of information utilities that makes access and use of the data simple and efficient. Three important developments make integration work:
    - The capability to plan, design, and execute activities and data management functions across the enterprise. Again using CE as an example, firms committed to CE have made

a big investment in modeling technologies because modeling provides a concrete, comprehensive vision of what needs to be done to the enterprise team, its organization, and manufacturing people. Within the enterprise concept, many of the decisions formerly in the domain of individual engineers and managers are made by an enterprise team of engineers, designers, and process and manufacturing engineers. Solids modeling provides a way of precisely defining a description of requirements of what needs to be done and by whom in terms that are understood by all members of the team. It also provides a means to demonstrate the need and the effectiveness of the proposed solution, without the cost of physical prototypes or actual implementation, since most of the models can be analyzed, tested, and executed through simulation.

- The technical foundation for the management of an entire enterprise with fewer obstacles. The use of such systems as CAD and electronic data interchange (EDI) from original equipment manufacturers (OEMs) and their suppliers is hampered by the different standards used by the major OEMs. Suppliers building parts for more than one OEM are faced with CAD and CAM equipment incompatibilities. Emerging software developments, such as solids modeling, are making such issues transparent.

- Continuous improvement across the enterprise. Integrating information provides a way for management to evaluate not only individual designs, but also current processes, plan for changes to those processes, evaluate their impact, and after their implementation, control them.

Organizational Issues

Obviously, for such an enterprise to work effectively, multitiered issues about multitiered organizations have to be resolved. Each organization must support the enterprise, yet difficulties are compounded tremendously when cooperation among departments is sought on even the most basic issues.

- Coordination mechanisms have to be established. Since there are few to begin with, they will have to be created, modified, and refined until optimum interaction and response is achieved. Reporting structures are the backbone of the organization. Managers will act as facilitators as well as supervisors. Managers will also assume a liaison role, a communication link among groups or teams, traversing the enterprise.

- The reward system must be put in place. Rewards, and their system of distribution, have a deep and lasting effect on the amount of cooperation among individuals and groups. The enterprise must reward cooperative work equitably across the company's administrative and proprietary boundaries. A member of a design engineering team, for example, might be rewarded on the basis of his or her level of cooperation with manufacturing engineers. Inequities across these boundaries may have to be resolved.

- Reporting structures must be set up. It is important to reduce the number of reporting layers among the groups in the enterprise. Enterprise management must develop reporting and communication systems that link individuals and groups to common enterprise goals.

- Supporting culture is critical to promoting cooperation within the enterprise. It has two principal components: trust and a long-term perspective.

- Without trust, people within an enterprise are less willing to share information, and information in an enterprise is a critical commodity. The person controlling that commodity enjoys a certain degree of power because of it. That person is not likely to share the power if there is no assurance that the person receiving it is going to use it to support enterprise goals. Therefore, a culture of trust predicated on openness and effective communication is essential to the success of the enterprise. The technological component of effective communication exists today; it is the human mind-set that must change to make it work.

- In a visual engineering enterprise, certain phases of the product life cycle will require more time to complete than others. Managers, as facilitators with a systems perspective, will see a competitive edge emerge as products get to their markets much more quickly than before. The design process, for example, will likely take longer if concurrent engineering is part of the enterprise's business plan. With the design, engineering, manufacturing, support, etc., functions involved up front to determine the design characteristics and manufacturability of the product, the concept phase is going to be extended. However, the cost and time savings downstream will more than offset these increases, result in significant efficiency improvements throughout the production stream, and get the product to the customer more quickly. All these translate into a much improved bottom line.

People Issues

In a visual engineering environment, certain important people issues emerge that require careful consideration and planning. Each member of the cross-functional team may now instantly understand the concept, and therefore, will have input into potential changes for improvement. This evolves into interdisciplinary cooperation, enabled by the mastery of technical skills and training in the fundamental principles of the integrated enterprise.

Interdisciplinary cooperation. People operating in this environment are going to be working with people unlike themselves. Designers, for example, will find themselves communicating beyond their traditional boundaries with marketers, manufacturing engineers, process engineers, support engineers, and a host of others. This cross-functional cooperation will require an ability to listen, communicate, and compromise.

Technical skills. A mastery of technical skills for all job types in the visual enterprise is assumed. Beyond this, however, staff from all functions must be confident and communicate knowledge of their skills to others outside of their specific area of expertise and across the enterprise.

Training. People need to be trained in the fundamental principles of the integrated enterprise concept, and specifically in the concepts of visual engineering. Engineers in all disciplines will have to wear several hats that may be uncomfortable at first: coordinator, team member, communicator, and conflict resolver. They will have to learn how to negotiate and yield to compromise. They will have to think "integration" on an enterprise level. "My way or the highway" no longer works.

Making it Work

To form consensus throughout the enterprise and realize the benefits of visual engineering on both the local and enterprise levels, all organizations have to start with a carefully thought-out plan and build a strategy based on what the enterprise's customers and competition are doing.

The tools of visual engineering are developed to make their users more competitive. The use of solids modeling tools and team building will expedite problem-solving by communicating ideas, simulating them, and gaining consensus *before* the expense of physical prototypes or implementation is incurred. Integration is based on standards, and integration teams should adopt those standards widely supported by technology vendors to ensure compatibility, interchangeablitiy, and ease of upgrading tools such as engineering workstations. Teams must also ensure that good exchange standards are in place so reliable communication between systems is possible. Data bases present problems across functional boundaries because of the issues of proprietary data. Technology does not inhibit, it enables. But for visual engineering integration to succeed, the organizational and people issues have to be dealt with at the same time as technology adoption.

7

Best Practices

PATHWAY TO PROSPERITY

Forward bound companies recognize that product development excellence is the key to sustained success, and that time-to-market, quality, and cost are the benchmarks of product development excellence. The following product development-related case studies share best practices. Some of the studies are technology-based, others, process-based. They focus on business problems and offer various alternatives that companies have used to solve them. No one company has a single set of best practices for new product development. The case studies cited offer snapshots of various aspects of the new product development process.

> *Best practices are the incorporation of things gone right into process knowledge that is transferred to others in an understandable manner.*

Each of the five case studies deals with a different aspect of the new product development process:

- *Case study one—design communications.* This study looks at the classic problem of getting product and manufacturing engineers on a common communication medium that is readily accessible to all.

- *Case study two—free-form fabrication.* This study focuses on the reality of "desktop manufacturing," with a process overview of stereolithography.

- *Case study three—workstation order fulfillment.* In this study we look at a re-engineering project that drastically reduced the time required to acquire workstations for engineers.

- *Case study four—high-performance computing.* The power of integration is highlighted in this example, featuring the coupling of workstation, scalable parallel processors, and traditional supercomputers to create the engineering "virtual desktop."

- *Case study five—predictive engineering.* This final study features predictive engineering best practices, as well as a study of an automotive component design.

Case Study One—Design Communication

Design communication is critical to the success of new product development. The case study outlined in Table 7-1 highlights how one company used visual engineering to address its design communication problem.

The development of a design involves the design specification, the design concept, and the engineering drawing. Design communication deals with marketing, product and manufacturing development, production parts suppliers, process equipment suppliers, and senior management. Design specifications deal with packaging the product into the design envelope space and the environment in which the design must work. The environment is usually described in terms of "boundary conditions" such as temperature range, noise, maximum length, width, height, etc.

Development of a new product involves much communication among numerous stakeholders in the design process.

TABLE 7-1
Case Study One—Design Communication

	Company—Automotive Manufacturer
Stated problem	• Design communication difficulty among design, manufacturing, and management.
Business impact of problem	• Communication problems delayed decision time, resulting in delays for new product development.
Competitive assessment	• One major competitor collocated all functions of product development team. • Another major competitor used physical mockups of new design concepts to facilitate communication at key meetings.
Best practices deployed	• Standardized use of solids models to facilitate a common view of new product development alternatives.
Business benefits	• Cut decision and meeting time by 50% with common and shared visual engineering model. • Single model used for concept, CAE, computer-aided industrial design (CAID), CAD, and CAM design reviews.

Communication as a development tool. Product and manufacturing development communication is the representation of the design in an engineering drawing format. The representation is the basic language that the designer, product engineers, and manufacturing engineers use to communicate within the design process. It is a well-defined language that allows manufacturing engineers to understand the product engineers' design intent and how to best manufacture the part. The representation is typically a set of 2-D principal and auxiliary views, which allows someone to construct a mental 3-D image of the concept. The syntax is well-defined and requires special training to learn. The semantics of making engineering drawings are rigorous and interpreting them also requires extensive training.

> *Product engineers and manufacturing engineers look at, interpret, and communicate design data differently.*

The purpose of an engineering drawing is to let the design engineer define a concept that can be manufactured from the drawing. Historically, the design engineer and the manufacturing engineer have had trouble translating their product and process alternatives into a business scenario that is easily understood by program and financial management.

Design communication should include marketing, product, manufacturing, and senior management input. New product development concepts and alternatives were traditionally developed as 2-D wireframe data, drawing formats difficult to understand by anyone except the designers and engineers. This made design communication among people other than designers and engineers very difficult.

Manufacturing engineers must visualize the design as a 3-D drawing to understand how to manufacture the product. Additionally, they must provide reasonable estimates for the facilities and tooling required to manufacture the product in the forecasted volumes. Adding to the confusion is the spreadsheet-driven financial analysis that compares various product and process alternatives against baseline models. Communication problems are real, and increase the decision time on choosing from alternatives. Physical mockups and models facilitate the communication of various design alternatives, but add time and cost to the overall process.

> *Design communication can be a major impediment to time-based product development if no one can understand the engineers' design.*

Design communication, present and future. More and more companies that manufacture engineered products are recognizing the need to reduce time in their overall product development process. Unfortunately, too many of these companies throw money and technology at the problem without really understanding the time-based elements of the product development process.

Design communication is usually not viewed as a process for improvement. In this case study we explore how one company looked at design communication as a process to be improved.

Increased competition places a premium on new product development time. Such time-based demands force design engineers and manufacturing engineers to make decisions faster. Figure 7-1 shows the traditional representation of a design. The 3-D wire frame design includes confusing overlapping lines. When two-dimensional plots are developed, the design can be even more difficult to visualize. The need to mentally reassemble the views slows the process.

The new process, incorporating visual engineering concepts, takes the original 3-D design data and converts it into 3-D visual engineering renderings that are much more realistic, allowing decisions to be made in a fraction of the time of the current process. The new representation of the design, developed using this process, is shown in Figure 7-2. The representation of a design concept as a realistic entity allows the concept to be instantly recognized and understood by marketing, product design, manufacturing, suppliers, and senior management in a truly simultaneous process. The visual representation does not change the data used to define the concept. The concept is still represented by line segments, circles, and arcs defined in dimensions. Two representations of the concept in different dimensions must be used to develop or construct a 3-D representation of a part. Mathematical 3-D operations are applied to this 3-D part representation to produce a realistic image.

130 *The Virtual Engineer*

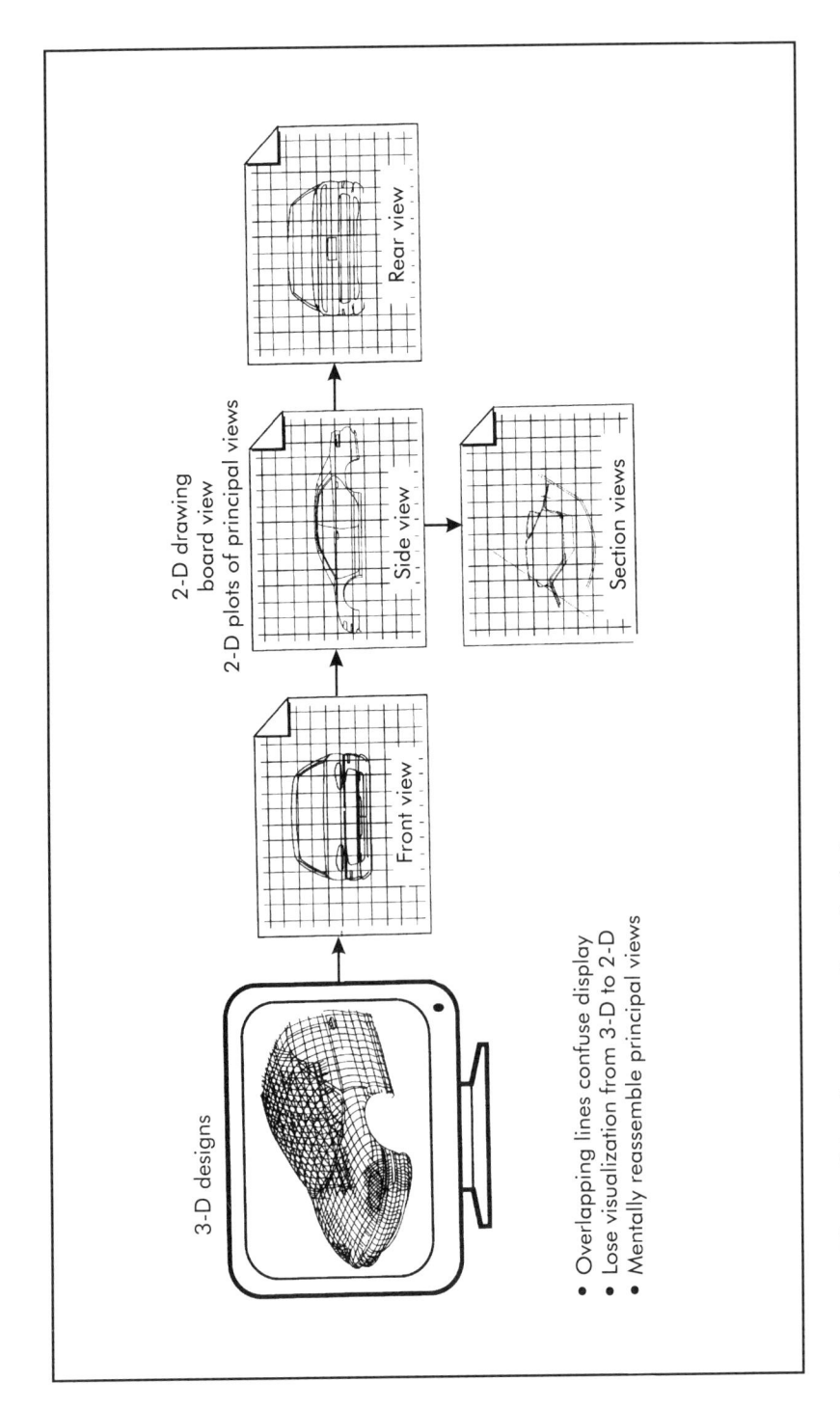

Figure 7-1. The traditional representation of a design.

Best Practices 131

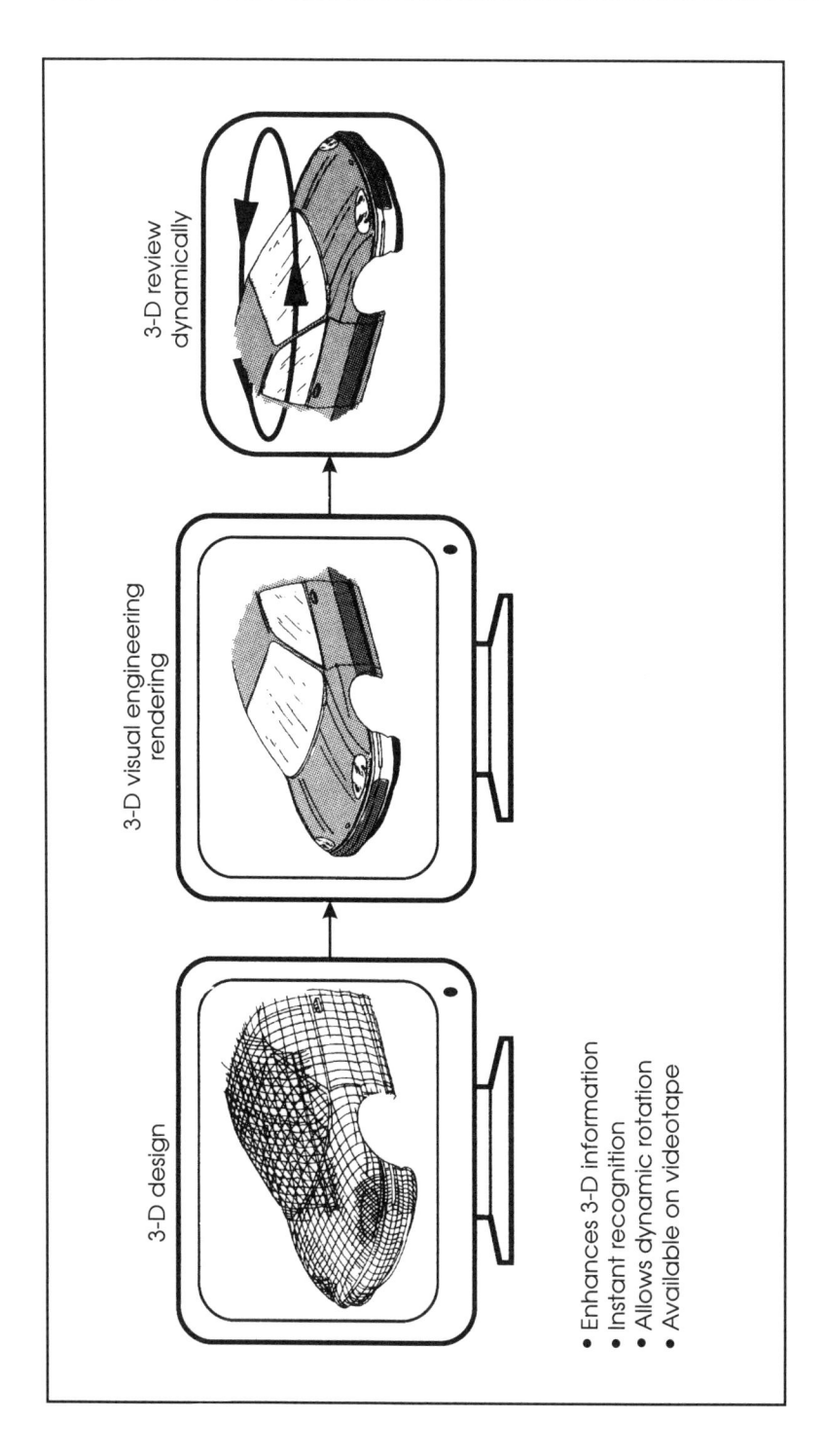

Figure 7-2. Representing a design using 3-D visual engineering.

Advances in data representation technology have accelerated the rendering speed of the design process. In the early 1980s, these redesigns would require 7 days of processing time (168 hours) on a $1 million machine to produce a single picture or frame. With the advent of engineering workstations, the cost and time to perform mathematical calculations has undergone incredible cost reductions and huge increases in speed. By the early 1990s, the time to produce the picture was reduced to 3 minutes and the cost of the workstation was closer to $90,000. Today, the frame can be produced in 1 minute on a $25,000 workstation. In seven years, a speed improvement of 10,000 to 1 with a simultaneous equipment cost reduction of 400 to 1 has occurred. Clearly, these revolutionary changes in technology produce dramatic changes in the process, particularly in the *speed* of the process.

The use of 3-D geometry has evolved into solids modeling, an enabling technology that promotes common dialog among product engineers, manufacturing engineers, and financial people. With 3-D capability, all involved are looking at the same mathematical description as a realistic rendering of the concept—a picture—so no time is wasted on interpretation of alternatives. In other words, an engineering representation can be accessed, viewed, and *understood* by nonengineers. This creates common visual engineering in which all stakeholders of the new product development process can take part without confusion.

> *Solids modeling forces a common visual engineering language among all process stakeholders on new product decisions.*

Concepts modeled in solids also can be used to perform predictive engineering without translation and incorporate automatic processes (such as automeshing) to prepare the concept for analysis. This enables the engineer to optimize the design in a very short time—hours or days—rather than weeks or months. Many alternatives can be evaluated before hardware prototypes are built, substantially reducing the cost and cycle time of the alternative evaluation process.

This speed improvement allows the engineer to answer a question before the question is forgotten. It reduces time between decisions and increases the number of times the engineer can ask "what if?", which is critical to improved and lasting enhancement of the process. A discussion of the advantages of predictive engineering appears in Chapter 5.

> *Design communication and understanding is greatly enhanced through the use of 3-D and solids modeling.*

Meetings between product and manufacturing engineers, suppliers, and senior managers are greatly improved through the use of 3-D and solids modeling. Productivity is greatly enhanced, and fewer, shorter meetings are required to reach consensus on new product design alternatives.

Faster decisions and fewer meetings increase engineering productivity, in addition to ensuring that new products are developed on time or earlier. Benefits to the business include improving the quality of the product, while simultaneously reducing the time and cost of new designs.

Case Study Two—Free-form Fabrication

The case study outlined in Table 7-2 highlights how another company used solids modeling to address its design validation problem.

Acme Manufacturing is a family-owned and privately-held, 100-year-old foundry with a reputation for high quality, slow delivery, and high cost. Part of this reputation is a result of the pride of craftsmanship of Acme's people, honed by generations of apprenticeships in tool and die making. Business had been flat for the last several years, mostly a result of multiyear contracts that require reductions of two to three percent per year in prices to the customer.

> *Acme Manufacturing, in business for over 100 years, had a reputation for high quality, slow delivery, and high cost.*

TABLE 7-2
Case Study Two—Free-form Fabrication

Acme Manufacturing Company—Automotive Parts Supplier of Cast Parts	
Stated problem	• Time, quality, and cost constraints in design validation for foundry tooling development due to traditional "craft-oriented" operating practice.
Business impact of problem	• Time delays and cost penalties associated with design validation and mold manufacturability.
Competitive assessment	• Competitors have faster turnaround and lower costs with comparable quality. • Newer, smaller suppliers installed "desktop manufacturing" at minimal cost to become a major competitor to traditional foundries.
Best practices deployed	• Use CAD-driven free-form fabrication techniques such as stereolithography and laminated object manufacturing.
Business benefits	• Direct transfer from CAD to CAM provides fast turnaround and exact design transfer. • Increased competitiveness due to improved quality, lower cost, and faster turnaround.

Acme has a third-generation family member taking over management of the company. Don is Acme's President, and unlike his father and grandfather, did not come up through the company ranks as an apprentice tool and die maker. Although knowledgeable and trained in the tool and die process, Don was sent to a leading Big 10 school where he majored in mechanical engineering, acquiring a personal interest in free-form fabrication and rapid prototyping. Don was concerned with Acme's eroding competitiveness in an era where speed and cost were the new benchmarks and quality was just assumed.

Don's educational background enabled him to understand, trust, and use mathematical simulation to perform computer-aided engineering to optimize a design before building physical prototypes. Recent computer price/performance changes and improved engineering software enabled even small models to accurately predict the response of a design in its functional environment at a cost even Acme could justify.

The linking of design with manufacturing convinced Don that technology was available to further improve his process. Don had invested in advanced CAE/CAD/CAM equipment for the past several years and was now ready to integrate free-form fabrication.

Acme was confronted with a rush request on a new braking system component for a major customer. One of the customer requirements was to use CAD data, package it, and then manufacture 250 prototypes for testing. The parts had to be ready in 60 days to meet the customer's new product cycle plans. Don was finally faced with an opportunity to fulfill his vision of using free-form fabrication in place of the traditional process.

Before investing heavily in the new equipment, he enlisted the help of a local service bureau to help speed up his mold design and validation process. Acme transferred CAD data to workstations at the service bureau. From the CAD data, Acme engineers worked closely with the service bureau to develop the new designs. Acme engineers used a workstation-based CAE/CAD/CAM system to create a CAD solids model of the new braking system component. This information was used to produce prototype parts directly.

Several alternative designs were generated within days. Working with the customer firm and its manufacturing engineers, Acme chose a design to optimize the customer's cost as well as significantly improve Acme's profit margin.

Once the decision was finalized, free-form fabrication, in the form of rapid prototyping, was necessary to reproduce the exact precision fluid-flow damping characteristics within the part. This precision was critical since it affected the performance of the overall braking system.

Without the use of solids modeling and free-form fabrication techniques, Acme would have reverted to traditional, craft-oriented practices. This would have made it difficult for Acme to meet the time-based needs of the customer. Also, much of the CAD-based data would have needed to be further translated into pattern drawings. In traditional pattern-making techniques, a skilled craftsman would take several days to translate the design into a physical model. Solids modeling, on the other hand, enabled the design engineer to create the product design in a day in the form of an electronic solids model.

Acme finished the CAD design and created a master model on the service bureau's laminated objected manufacturing (LOM) machine. This machine automatically builds parts from layers of paper cut with a computer-controlled laser driven by the CAD geometry. The resultant braking system component envelope was approximately 4-in. (10-cm) long by 3 in. (8 cm) in diameter. It took the LOM about 8 hours to build the model, which was as geometrically accurate as a final manufactured part.

Free-form fabrication techniques such as LOM and stereolithography are making CAD-driven desktop manufacturing a reality.

After completion of the prototype, Acme presented it to the customer for review and assessment. Inspection revealed that the CAD data transferred to Acme was not the latest engineering release, not uncommon with most part designs in any manufacturing company. New CAD data was provided, and Acme reproduced a new prototype only 1 day later. This LOM process of 1-day turnaround compares to a change and turnaround time of weeks using the wood models. It is clear that Acme's timing and cost would not be competitive without the CAD-driven, free-form fabrication technique.

After acceptance by the customer, Acme used the LOM part to develop patterns for sand casting. Acme cast 250 copies of the braking system component out of an aluminum alloy. These parts were then shipped to the customer to be used in vehicle testing. The entire project took approximately 5 weeks, counting the delay for engineering changes and reinspection. Using the traditional process would have required a minimum of 4 to 5 months.

In addition to the obvious benefits of creating prototypes which exactly matched the design, Acme estimated the free-form fabrication technique saved about 3 weeks in the tooling process. Acme's customer was clearly pleased with the results, especially the fast turnaround on the engineering change problem.

> *Free-form fabrication techniques saved about 3 weeks in the tooling process, compared with traditional pattern-making practices.*

Acme's use of the geometric data that was transferred directly to them over phone lines allowed them to jump-start the design process. Once the braking system component geometry was isolated from other surrounding components, the design was quickly meshed for analysis on how best to interpret its geometry for manufacturing. Figures 7-3a and 7-3b are mesh and solid representations of the braking system component.

Visual engineering was used to quickly validate not only the design of the part, but its packaging and interference fit with the surrounding components of the braking system, tire and wheel assembly, and suspension system of the vehicle. Figure 7-4 shows the *visual engineering* representation that resulted from the CAE to CAD/CAM through LOM process.

138 *The Virtual Engineer*

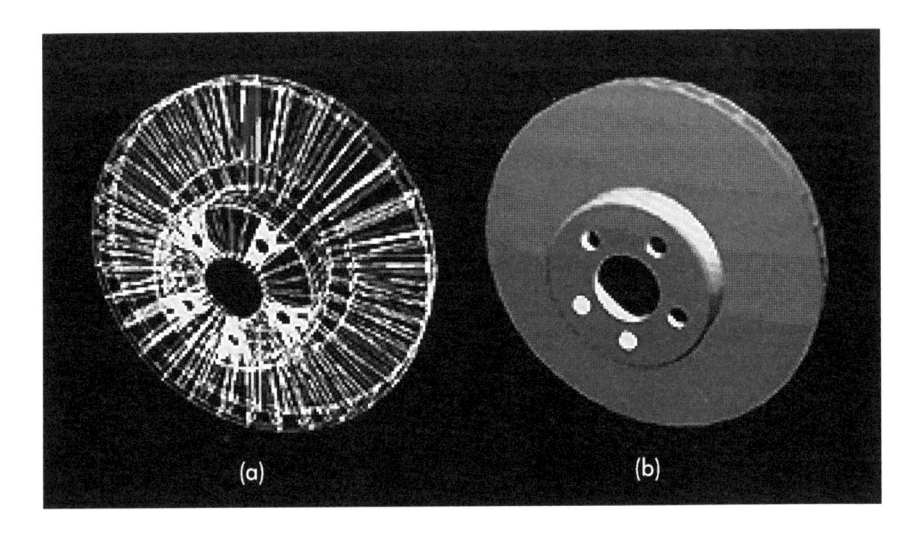

Figure 7-3. The mesh representation of a solids model (a). The solids model itself (b).

Figure 7-4. Validating the part's fit with the entire assembly using visual engineering.

Acme's use of the LOM process is not unique. LOM technology uses a single laser beam and thin, roller-fed sheet material to create complex solid objects. The sheets, which are precoated with heat-sensitive adhesives, are laminated one on top of the other to create a multilaminar structure. Once a layer is bonded, the CO_2 laser cuts the outline of the specific cross-section that represents one of the thin cross-sections within the 3-D object. This process continues until all layers are cut and laminated, ultimately creating a solids model of the CAD part.

LOM is used to validate not only the design of the part, but whether the part can be manufactured. For components that must nest within a packaging envelope of other components, LOM is a tool for making parts quickly to validate packaging on prototype designs.

An axiom in the automotive industry is that the supplier that makes the prototype has a much greater chance of winning the ongoing production award for the parts.

Free-form fabrication and CAD have given new life to Acme. Don has not only realized his dream of rapid prototyping, but he is in the process of developing and installing Acme's own service bureau capability. Acme has used free-form fabrication techniques to increase competitiveness through higher quality (part exactly matches the design), lower cost (tooling process was reduced by more than 50 percent), and faster response (days instead of weeks).

Acme is one of many companies that have successfully deployed computer-based technologies and free-form fabrication techniques to increase overall competitiveness. These new processing technologies have not just replaced the skills of the eroding base of pattern makers, but have actually embedded their craft skill base into repeatable computer software programs.

Acme is a company that should not only survive the continued reduction of suppliers, but flourish with its use of CAE/CAD/CAM.

Acme could never have met the customer requirements through traditional pattern-making techniques. Don deserves credit for using free-form fabrication. He created *virtual* pattern makers by integrating free-form fabrication with design. With the continued price/performance breakthroughs in computer-aided technologies, free-form fabrication techniques, such as LOM and stereolithography will become new product development best practices.

Case Study Three—Workstation Order Fulfillment

New product development in the automotive industry is time-based. New programs typically take about 3 to 4 years and cost several billions of dollars to launch. Studies have shown that delays of popular new products cost about $4 million per day in lost profits. Because this lost profit contribution is so staggering, companies explore all avenues to improve the process so that days and weeks can be cut.

Timely order fulfillment of tools for the engineer, as outlined in Table 7-3, is critical to any new product development program.

> *Going from identification of a need for an engineering workstation to actual receipt and use on the desktop took an incredibly long 30-40 weeks.*

Our focus company is a $10 billion automotive components manufacturer that purchased about 300 to 400 workstations annually for its engineers to use in new product development and related activities. Traditionally, most engineering workstations were bought for small local projects scattered among the various engineering functions, plants, and divisions. Realizing that the process of new order fulfillment was taking too long, management of the company commissioned a study of the purchasing of engineering workstations to better understand the problem and develop alternatives for cutting time from the overall process.

TABLE 7-3
Case Study Three — Workstation Order Fulfillment

Company — Automotive Manufacturer	
Stated problem	• Took too long to get an engineering workstation with an excessive order fulfillment process that required about 30 to 40 weeks.
Business impact of problem	• Lack of timely order fulfillment of engineering workstations was causing time, cost, and quality impediments to new product development.
Competitive assessment	• One competitor lets engineers handle procurement, thus slashing administrative paperwork and time in the system. • Another competitor outsourced a technology refresh contract to a major workstation supplier, thus getting out of the order fulfillment business.
Best practices deployed	• Re-engineered order fulfillment process and cut delivery time of engineering workstations from 30 to 40 weeks to 7 days or less.
Business benefits	• Engineers have tools fast and no longer have to acquire system with nonvalue-added activities. • Company reduced workstation prices by over 10% while improving cash flow by over $20 million in 1994 alone, an increase of 10%.

Interviews were conducted with both internal and external stakeholders of the order fulfillment process to assess the current process and identify things that had gone wrong.

These interviews revealed the following:

- An excessively long order fulfillment process that averaged 30 to 40 weeks after identification of needs.

- Confusion among requisitioners, buyers, and suppliers about their roles and responsibilities.

- Inability to keep up with the rapid price/performance changes of the computer industry.

- Lack of a vision for future workstation requirements.

- Difficulty in managing the complexity of supplier product and technology offerings.

- Excessive nonvalue-added time, as engineers focused on expediting orders through the system rather than on designing parts.

- Absence of a process to rapidly deploy newly-released technologies to the desktop of the engineer.

- Management nightmares and nonvalue-added time in tracking desktop assets while simultaneously ensuring latest technology.

The number of nonvalue-added steps and an excessive amount of paper created in the process presented a re-engineering opportunity.

The first phase of the study identified an excessively long order fulfillment process of about 30 to 40 weeks for an order that had no changes or clerical errors. Figure 7-5 shows the major steps of order fulfillment. The seven order fulfillment steps covered the point from when the system was specified until it was installed and paid for.

Breakout groups identified root causes of the problem. After a list of the causes was captured, another group focused on ways to improve the process of order fulfillment of engineering workstations. This resulted in a process reduction from 30 to 40 weeks to 16 to 18 weeks, or a 50 percent improvement. However, this plan was rejected!

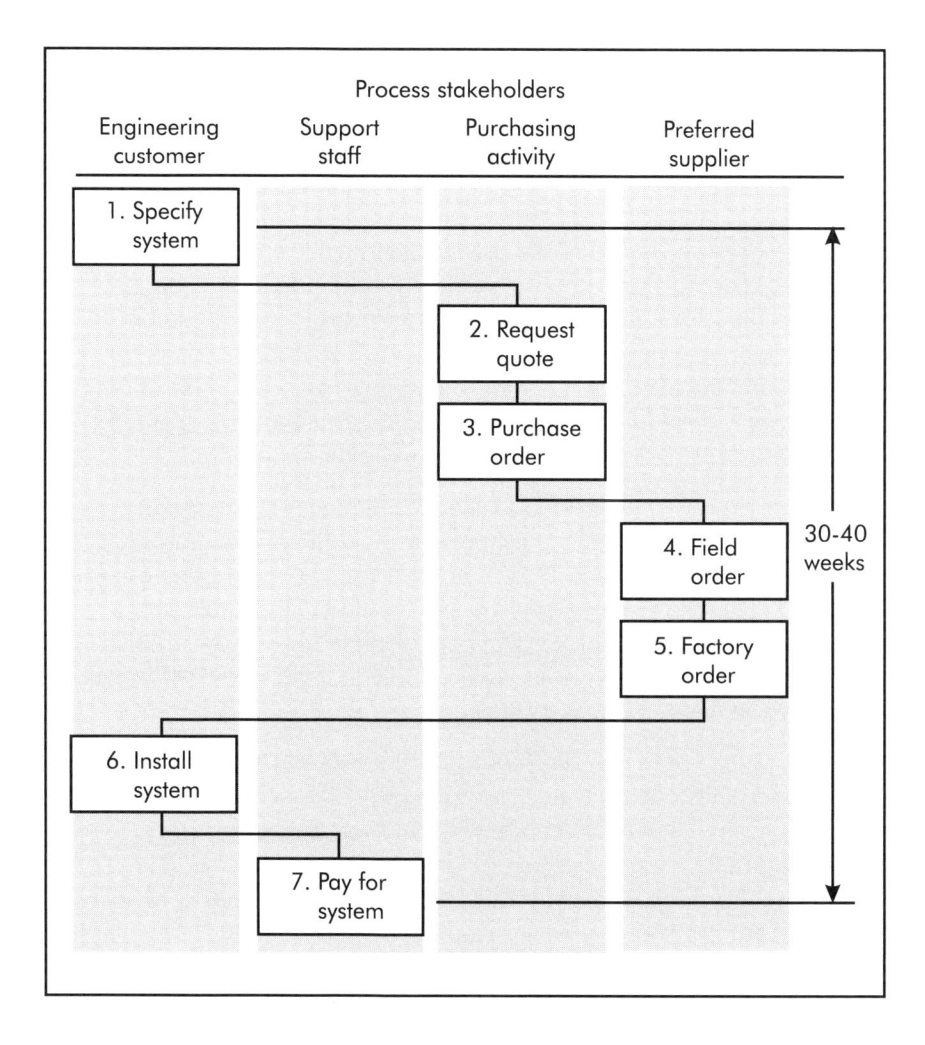

Figure 7-5. Existing order fulfillment process for engineering workstations.

An executive at this company looked at Dell Computer, which at about this time (early 1993), had been establishing the benchmark for fulfillment of phone orders of personal computers. If a customer could call Dell's 1-800 number, and after a few minutes, place an order and be assured of delivery within 48 hours, why couldn't engineering workstations, though a little more complicated to purchase and build than personal computers, be received in 7 days or less?

> *Going from 30 to 40 weeks for workstation order fulfillment to 7 days or less seemed like an impossible task to the team.*

The executive champion gave the team another challenge—come back in 60 days with a strategy and plan for 7-day delivery. Traditional process improvement was a great change from the horrendous baseline of 30 to 40 weeks, but still not good enough to meet the time-based needs of the engineer. To achieve the envisioned improvements, drastic changes to the current process needed to be made.

The team quickly realized that the old, serial way of doing business would not work, particularly in light of the 7-day goal. Copying a practice from the production side of the business, the team went out to try to capture future requirements for engineering workstations. The number totaled over 1500 for the next 5-year period, and over 1000 for the shorter and more reliable next 3-year period.

The next task froze the technology offerings in the form of standard specifications, or "rapid specs." Benchmarking played a big part in this step as various workstations were tested against a wide variety of desktop applications. Although technology was continuously changing, it could be frozen in 90- to 180-day intervals.

> *The financial community in most companies has no interest in re-engineering business processes that take control away from the bean counters.*

After deciding upon classes of machines, a project was approved for about $30 million to cover the forecasted 3-year workstation requirements of the engineers. Due to the rapidly changing technology of engineering workstations, management empowered a cross-functional team to make procurement decisions for the company as a whole. This team included people from Purchasing, end-users, and systems support personnel from various areas of the company.

The workstation supplier was an integral part of the re-engineering team. In visits to the supplier's factory, it became evident that 18 minutes of direct labor assembly and over 9 hours of test also needed re-engineering. The supplier appeared to not be in the workstation business, but the material handling business. Workstation components were shipped from all over the world to this factory for assembly and test. Current systems to support this operation did not meet the 7-day delivery vision.

Seven-day delivery meant that the workstation supplier had to stage units near the customer and be ready to ship on a moment's notice.

The re-engineered process called for 7-day delivery, not unlike Dell's 48-hour delivery. Since the value chain for workstations was about 6 months from silicon to engineer's desktop, some changes had to be made. In light of the forecast and rapid specs, achieving 7-day delivery was the same as giving the supplier a blanket order for workstations.

The supplier converted its transportation agent's facilities into a customer delivery center. This center stored and staged workstations in anticipation of its scheduled request from the customer. This float created the inventory needed to satisfy the 7-day delivery of workstations to the engineer's desk. Having a local center also provided instantaneous communications.

After launching 7-day delivery of engineering workstations, the team established a process and developed best practices. Figure 7-6 shows the seven major steps of the re-engineered process.

Today, the re-engineered process has evolved to one where the supplier manages the engineer's desktop. The supplier is paid a monthly lease fee to ensure that the engineer has the best technology on the desktop. Trust is an integral part of this arrangement, and the contract goes up for bids every 3 years to ensure competitive pricing.

The 7-day vision has resulted in engineers never having to worry about desktop tools. The supplier is responsible for benchmarking and certification of new technology as well

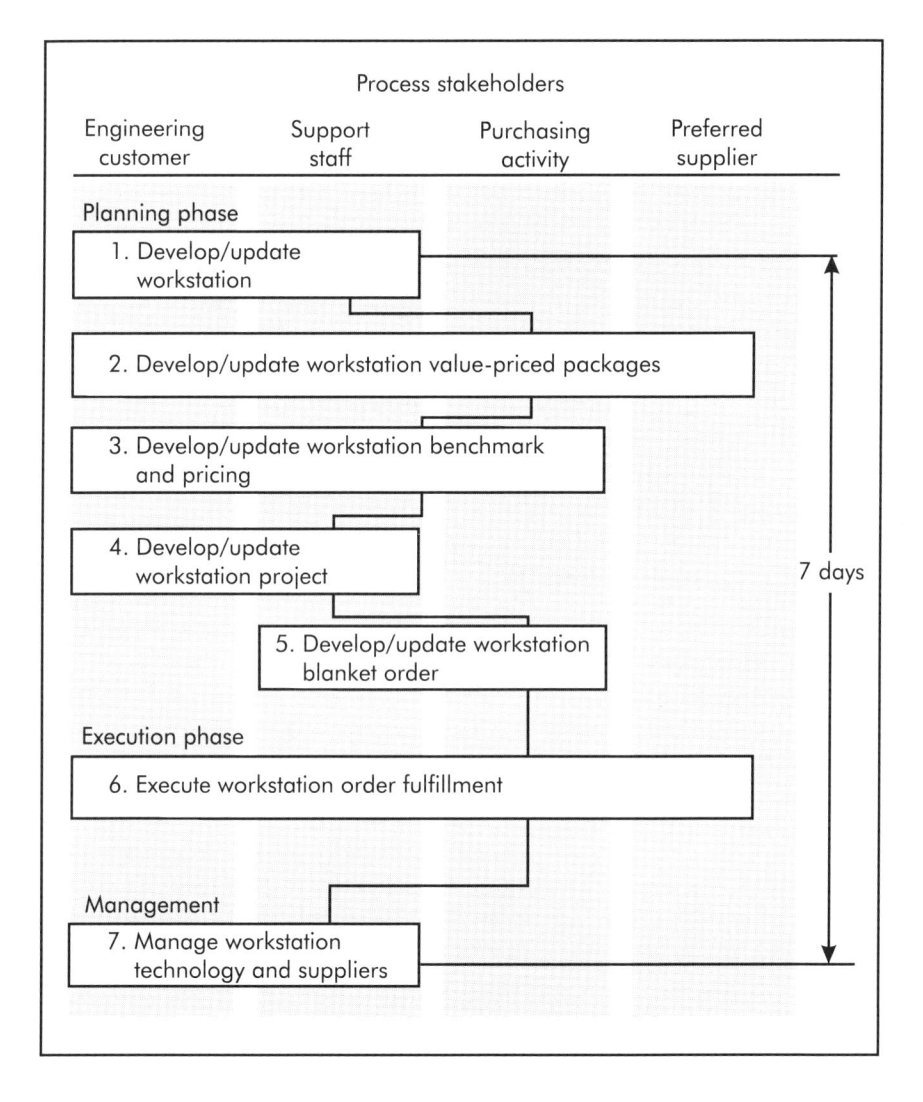

Figure 7-6. Envisioned order fulfillment process for engineering workstations.

as continual upgrading in support of the engineer's needs. This arrangement takes advantage of the core competencies of both companies.

> *Timely order fulfillment of engineering desktop technologies is critical to the success of time-based product development.*

This case study is unique, but not for its re-engineering of workstation order fulfillment. It puts focus on a malady that affects many companies today in their pursuit of computer-based technologies. Many companies put some of their best computer systems personnel or engineers into roles of expediters trying to work around the system just to get tools to the engineers. As this case study showed, it is often easier to outsource a process instead of going though the internal politics involved in changing or improving it.

The benefits of this re-engineering effort were beyond all expectations of management. In addition to meeting the fundamental objective of fast delivery of workstations to the desktop of the engineer (7 days or less), there were financial benefits as well:

- Integrating future requirements and standardizing classes of machines allowed the company to achieve an additional discount that resulted in a savings of over $2.4 million over the 3-year period.

- Standardized lease arrangements through a blanket lease order improved the cash flow savings from $7 million to $17 million during the course of the project, an improvement of $10 million.

- Elimination of nonvalue-added activities throughout the process effectively reduced headcount, generating an annual cost savings of more than $1 million.

- Most importantly, engineers had their tools quickly and at the time they needed them. Fast delivery improved productivity. In addition, there were attendant improvements in cost, quality, and timing to their particular new product development programs.

Case Study Four—High-performance Computing

Computational capability is essential to 21st century product design. The higher the level of computing performance, the greater the edge in the marketplace. Table 7-4 highlights a case study addressing this issue.

All new product development is managed to time-based schedules and milestone deadlines. Many of these milestones are driven by prototype reviews to assess product design completeness and process capability.

Traditionally, new engineering requirements are validated through the design, build, and test of physical prototypes. Today, government-regulated safety requirements have increased in the automotive industry by more than 25 percent. Many of the new safety regulations require extensive prototype development and design validation. Creating these prototypes is time-consuming and costly, so companies are increasingly using computer-based models to simulate a wide variety of operating conditions for new designs. Physical prototypes, if built at all, are used only to validate the computer-based model and design. Today, companies in the automotive industry increasingly rely on computer-based models and simulations to meet engineering requirements, while simultaneously maintaining design schedules.

> *One automotive company has estimated that its prototype costs for new products is more than one-third of the total vehicle program investment.*

In the automotive industry, a typical new vehicle product development program requires an estimated $3 billion investment, with about 33 percent or $1 billion of this new

TABLE 7-4
Case Study Four—High-performance Computing

Company—Automotive Manufacturer	
Stated problem	• Time-to-market pressures on new product development require more computational resources to meet cost, quality, and time objectives.
Business impact of problem	• More than 75% of a new vehicle's manufacturing cost is sunk at time of engineering release.
Competitive assessment	• One competitor established collocated teams that handled all new design, development, and launch functions of a new vehicle. • Another competitor dedicated traditional supercomputers to key engineers faced with high-performance computing needs.
Best practices deployed	• Implemented mixture of traditional supercomputers, scalable parallel processors, and workstation into a virtual network for new product development.
Business benefits	• Engineers have all collocated CAE resources to vehicle platform teams for local management. • Company has increased CAE compute capability 100 times at a reduced cost compared to previous practices.

product development investment spent on prototypes. Reducing the number of prototypes built by 50 percent represents a cost avoidance opportunity of about $500 million.

This case study shows how a vehicle development team implemented their own localized CAE resource to allow predictive engineering to be performed in minutes or hours instead of days or weeks. Prior to local CAE capability, the vehicle team had to rely on centralized (such as Cray C-90s) resources for their high-performance computing.

The proprietary software and operating mode of these supercomputers made data management and file sharing with the desktop workstations of the vehicle team engineers impossible. Many of the analysis jobs were small enough that they could probably have been run more efficiently on scalable parallel processors (SPPs) or workstations. Now, with control of their own CAE resource utility, engineers can manage and control their job runs.

Team members demanded localized and immediate access to predictive engineering capabilities, paramount for enabling engineers to iterate design concepts quickly.

The current process uses a dedicated supercomputer that allows the engineers to perform CAE in a time-sharing, batch-run mode. The objectives of the Management Information Systems (MIS) department were to minimize the cost of computer hardware and to replace manual processes with computer programs that performed the same functions.

> *The MIS (provider) operating objectives did not match the needs of the product development engineers (customers).*

The engineers' objective was to develop a design that incorporated the best quality while reducing cost to manufacture and time to develop. The MIS (provider) objectives did not match the engineers' (customer) needs. This misalignment of objectives is not uncommon in industry, and is particularly detrimental to engineering productivity when it applies to new product development systems.

In developing a new strategy and plan, engineers were interviewed to ascertain their requirements for existing and envisioned predictive engineering capabilities that would replace hardware prototype evaluations. The interviews were enlightening and revealed many flaws in the way the company was currently doing business.

> *The Cray C-90s were being used 100 percent of the time, although over 70 percent of their work load could have been run on workstations and scalable parallel processors.*

- Many jobs that required less than 5 minutes of analysis actually waited in job queues at the supercomputer for several days at a time.

- Often after waiting for a day or so, a large simulation would crash due to an input or modeling error, thus requiring the process to start over from the beginning.

- Engineers tended to load up the system on Monday and Tuesday of the week. This resulted in an oversubscription of memory on the Cray C-90s, thus putting their jobs in queue waiting for their turn.

- One large simulation could tie up the supercomputer at the expense of many smaller simulations that could have been run instead.

- No network management system existed that would query jobs and send them to available processors.

Interviews with the engineers (customers) resulted in the development of a pilot for an envisioned CAE resource that would exceed the needs of the engineers. This pilot resulted in a CAE resource plan that would put localized management and control back into the hands of the engineers. The CAE resource plan for the vehicle platform team was a combination of workstations, SPPs, and time-sharing on the Cray C-90s. A typical new vehicle development program has about 300 engineers, and all of them need access to a dedicated engineering workstation on their desktop.

> *The combination and scalability of Convex SPPs and HP workstations created a virtual desktop engineering center.*

Studies, as a result of interviewing the engineers, showed that over 70 percent of the analysis on a new vehicle could be done either on workstations or SPPs. Although the analysis would take longer on an engineering workstation than a Cray C-90, *the total turnaround time* on an engineering workstation would be substantially less.

In this planned CAE resource pilot, RISC-based workstations were used because of their industry-standard UNIX operating system, stability of company technology, and workstation scalability up to supercomputers. Figure 7-7 shows the integration of RISC-based workstations, dedicated supercomputers, and high-speed communications into a *virtual engineering desktop*. The workstation supplier was contracted to manage the virtual desktop and ensure that it was continually upgraded to the latest technology available to the engineer.

Matching of the provider (MIS) objectives to the customer (vehicle development engineers) objectives resulted in:

- Improved quality. Increased use of visual engineering, possible on the new workstations, coupled with representative mathematical simulations promoted more accurate analysis that increased quality and design robustness.

- Fast turnaround. Engineers no longer had to wait for their computational resource answers. The CAE resource for predictive engineering was local, on-line, and on-demand.

- Accurate results. Engineers had control of the CAE resources directly from their desktop, reducing the time between concept development and concept validation. This increased how often the engineer could ask "what if?," resulting in more accurate models that were used to improve the quality of the final design.

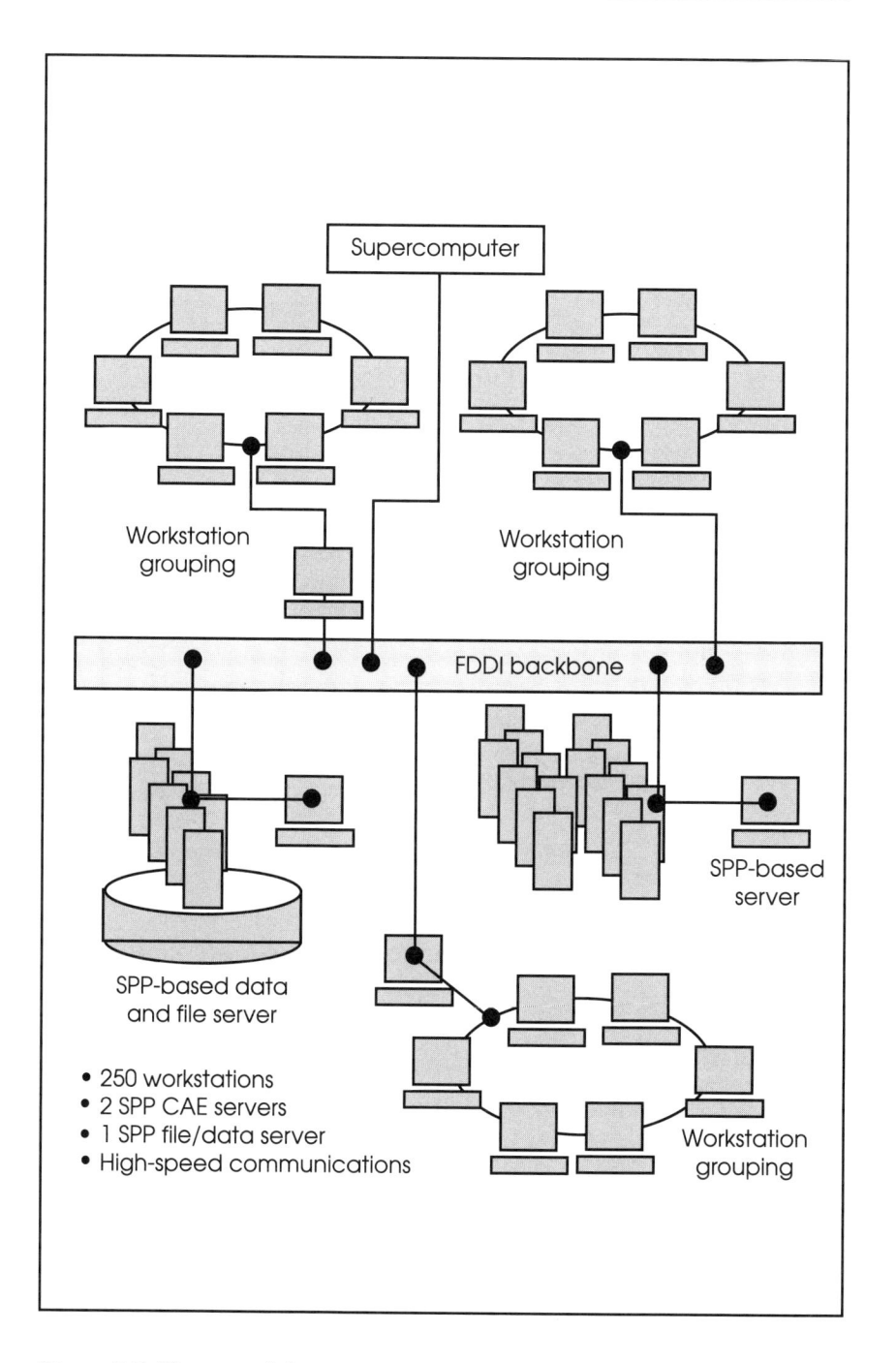

Figure 7-7. The network becomes a virtual engineering desktop.

- Large models for total vehicles. The CAE resource developed and analyzed large vehicle models in a much faster turnaround time, allowing predictive engineering to drive the process throughout its cycle.

Case Study Five—Predictive Engineering

Predictive engineering is a new design concept applicable to mechanical systems, subsystems, and especially components. Table 7-5 highlights how predictive engineering can be used to meet quality, cost, and time objectives of new parts and systems.

TABLE 7-5
Case Study Five—Predictive Engineering

Company—Major Automotive Manufacturer	
Stated problem	• Existing design process and practices do not allow enough time to meet quality, cost, and timing objectives for new products.
Business impact	• Parts and systems had to be released due to program timing constraints without having met the design targets for quality and cost.
Competitive asessment	• Another competitor used dedicated workstations with an outsourced engineering design firm to accelerate time in the process.
Best practices deployed	• Standardized use of predictive engineering to show a systems optimization first, and then a component optimization next.
Business benefits	• Reduced time in the process to allow for more analysis and test alternatives to reduce prototypes and meet quality and cost objectives. • Single model used for concept, CAE, computer-aided industrial design (CAID), CAD, and CAM design reviews.

As mentioned earlier, on a typical new vehicle development program, about one-third of the overall investment is for physical prototypes used to validate various design alternatives. Total cost of these prototypes generally exceeds $1 billion. This study focuses on using predictive engineering to convert many of the initial physical prototypes into computer prototypes.

The traditional process begins with the engineer describing the concept to the CAD designer. The designer begins the design from scratch or, more likely, retrieves an existing design similar to the description given by the engineer. The design is usually stored on a disk and is retrieved and modified to "fit" the current concept. Figure 7-8 depicts this traditional process.

A primary objective is to make sure the design "packages" correctly without any interferences from surrounding parts. After the engineer and draftsman are satisfied that the fit is not a problem, the design is transmitted electronically to another engineer using an engineering workstation.

The engineer then performs analyses to make sure the design does not fail. After the design is approved, a physical prototype is made and tested for final validation. The process seems simple, but requires separate disciplines to accomplish the functions of the design engineer, designer, analyst, and the manufacturing engineer. In addition, this process takes days and months to complete. The new process, as shown in Figure 7-9, is changed by eliminating the need for two disciplines, the designer and analyst.

In the new process, the design engineer performs the predictive engineering from an accurate, electronic, 3-D sketch in the form of a solids model. This was not possible 10 years ago and represents the new capability of the 21st century. The engineering drawing is replaced by a solid, 3-D representation of the part.

Meshes for performing analysis can be added automatically in minutes or hours rather than days and months. By having supercomputers available on demand, the engineer obtains predictive engineering results in hours or days.

156 *The Virtual Engineer*

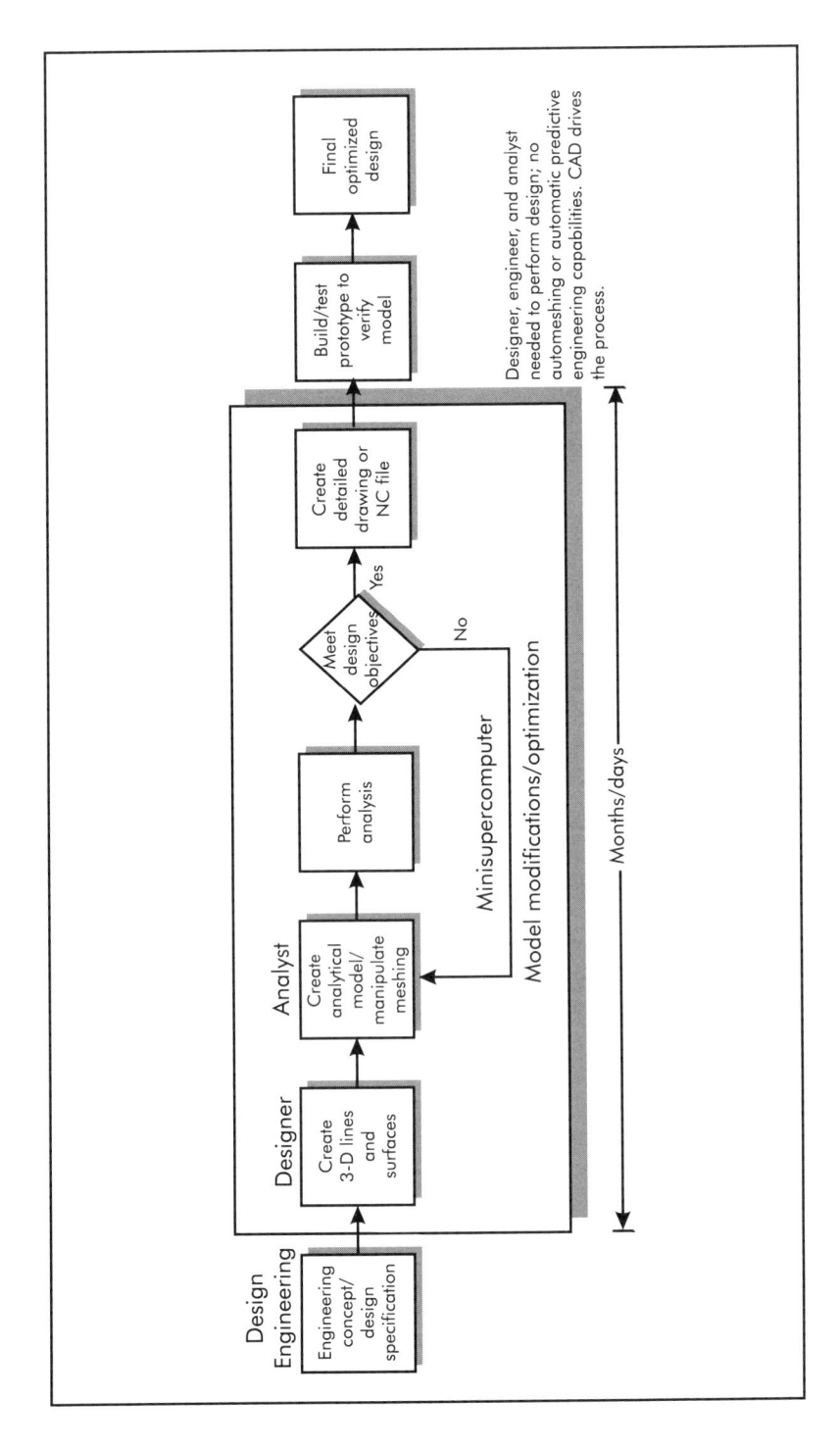

Figure 7-8. Traditional design process.

Best Practices 157

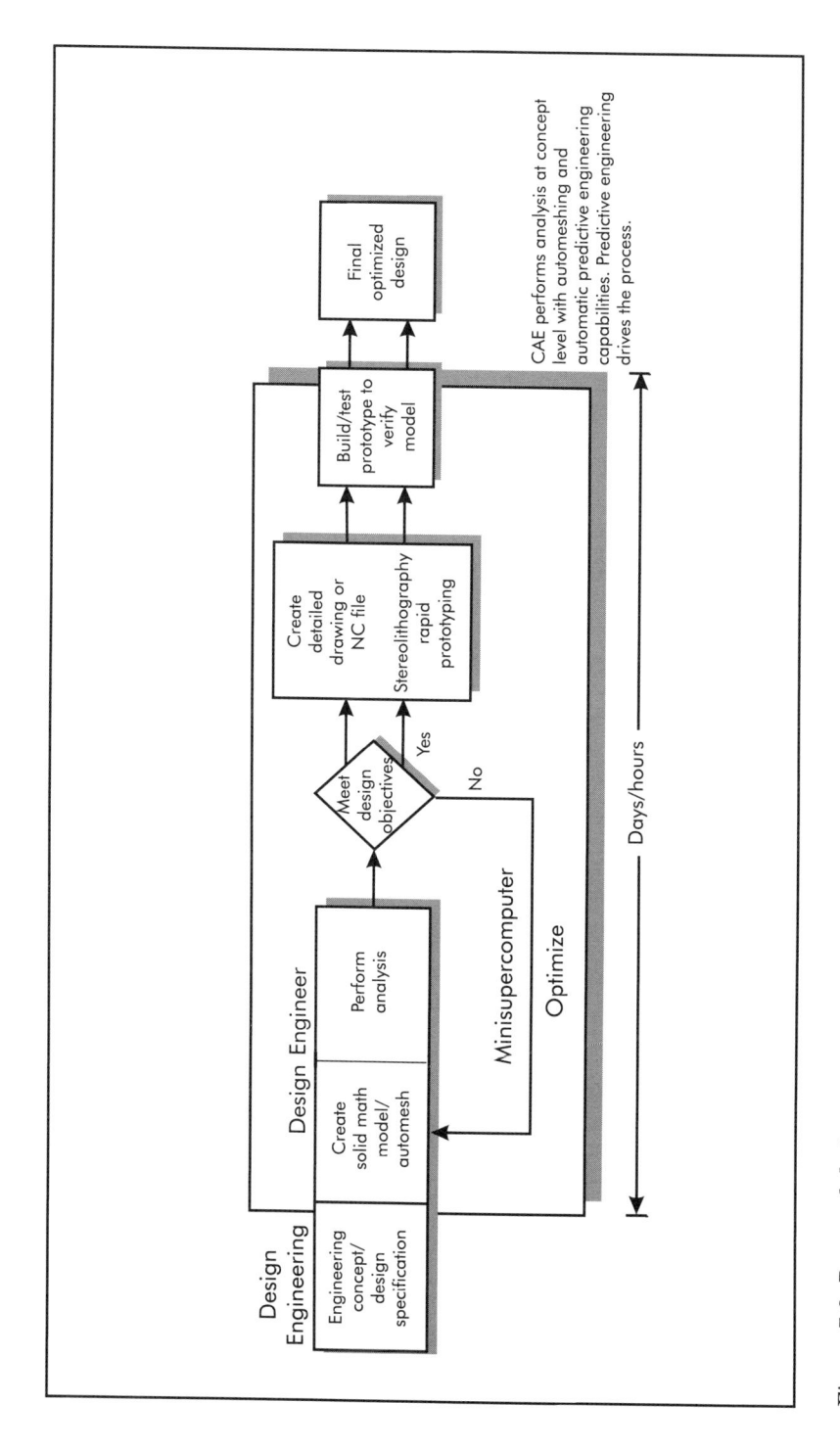

Figure 7-9. Proposed design process.

Because results are acquired fast, many more design alternatives can be evaluated for improved quality in the same or less time than before.

Using predictive engineering, taking minutes or days, rather than building physical prototypes, eliminates significant time in the design cycle and reduces the number of prototypes, which are very expensive in terms of dollars and time. Rapid prototyping allows CAE and CAD models to be manufactured without translation immediately after the engineer and management accept the concept. The part can be made overnight and returned to the engineer within hours.

Predictive engineering was used on a chassis steering system component on a new vehicle. This chassis component was a complex design, as measured by its past warranty costs, difficulty to manufacture and assemble, excessive weight, and packaging envelope requirements with surrounding subsystems in the vehicle.

For the re-engineered component, solids modeling was used to establish a common and shared geometry model throughout the design and development process. This shared geometry assured fast reaction to design and manufacturing requirements.

> *A common and shared geometry model was critical for establishing data translation among the various CAE, CAD, and CAM programs.*

Predictive engineering is an enabling design and development philosophy that uses CAE to predict the performance of designs under varying conditions. The steering component's input design data, in the form of line and surface geometry, was transferred directly (no master translation) into the solids modeling system. This system could create solids from the reference geometry. This step was critical to the predictive engineering process since the geometry was shared from the design master to the solids modeling system to the physical prototype.

Once the solid model geometry was generated, the product development engineer performed preliminary package studies to look for interference with surrounding system components and to establish envelope geometry. Next, the product development engineer conducted mass property calculations and kinematic analyses.

A key aspect of predictive engineering is the ability to quickly and accurately generate CAE analysis models to meet overall vehicle design targets for stress, crash, computational fluid dynamics (CFD), and noise, vibration, and harshness (NVH).

Predictive engineering allows CAE analysis models to be developed quickly and more accurately than traditional forms of analysis and test.

The key to fast and accurate CAE analysis is robust automatic mesh generation capabilities within the solids modeling software. Based on the analysis results, the solids model is modified and a new CAE model is constructed, analyzed, and evaluated by the design engineer. This process is repeated by the design engineer until all original design objectives have been satisfied, or to meet changing customer requirements, design targets, or analysis boundary conditions.

The use of solids modeling promotes dialog and communication between the various component stakeholders responsible for the steering system design. The solids-based analytical model predicts the behavior of a particular steering system component design concept before prototype hardware is ever built. This is the iterative process of predictive engineering that slashes time from the traditional design and test process used previously on the steering component design. The previous design contained more than 18 variants of design to meet all vehicle model needs.

Predictive engineering allows analysis, test, and validation of design concepts before prototype hardware is ever built.

The cost of the existing steering system component was about $100 and the part weighed about 20 lb (9 kg), fully assembled. The current warranty cost of the component was about $20 per unit, or about 20 percent of its total cost. Engineering costs, including facilities, tooling, and prototypes were estimated to be about $20 per unit.

Design objectives were very aggressive and included a target cost reduction of more than 25 percent, with specific objectives set at reducing engineering and warranty cost by more than 50 percent.

Predictive engineering was used to meet the objectives set out for the redesign of the steering components. An existing CAD design was used as the beginning baseline. Rather than continuing the design in the traditional CAD representation, it was redone in a solids format. Changes were made in this representation to achieve the initial design, which allowed packaging of the component within the design environment. Part redraw time in solids took 2 days, plus one-half day for packaging modifications. Traditional CAD would have required 4 to 5 days.

> *Predictive engineering allows analysis, test, and validation of design concepts in 50 percent of the time it would take for traditional CAD.*

A 50-percent reduction in time was achieved, including a complete redefinition of the part in solids. The engineer immediately began to perform predictive engineering. Changing the model required 4 to 6 hours, and preparing the model (automeshing) for analysis took about 45 minutes. Performing the analysis and obtaining the result for review took about 1-1/2 hours.

Using the traditional CAD approach would have required 2 days for a change, about 4 to 5 days to mesh, and 3 to 4 days to obtain the results for review. The differences in time requirements between the traditional CAD-based approach to design and predictive engineering are tremendous. Table 7-6 details the changes per function and total time differences between traditional CAD and predictive engineering.

The time estimates of Table 7-6 are based upon an actual case study and represent a reduction in time of four or five to one for obtaining a predictive result. It does not include design iterations to improve quality or cost, or the physical prototype to test for final functionality. The design iterations vary depending upon the experience level of the user, the complexity of design, and the time available for evaluating design alternatives.

TABLE 7-6
Task Completion Time Comparison—Traditional versus New

Design Function	Traditional CAD (hours)	Predictive Engineering (hours)
Develop concept	N.A.	16
Packaging	32-40	4
Design change	16	4-6
Design preparation	32-40	3/4
Design review	24-32	1-1/2
Total time	104-128	24-35

If the engineer performs design iterations, the time required using traditional CAD/CAE is 720 to 880 hours, while predictive engineering requires only 60 to 80 hours. This represents a 10 to 1 improvement in the design process alone. After the design phase is complete, a physical model must be built and tested.

Physical prototyping, using traditional methods from CAD data, requires 4 to 6 weeks at a cost of $150,000. Rapid prototyping of the design can be accomplished in 4 days at a cost of $14,000. Using predictive engineering and rapid prototyping can achieve reductions of 10 to 1 for the total design cycle. It should be noted that time reductions in getting new products to market can be as much as 50 percent.

Recent articles in industry and trade publications have indicated that the Japanese can develop a new car in 24 months. Changing the process that U.S. companies deploy would allow them to leapfrog foreign competition without a sacrifice in quality or cost. In practice, the quality of the product would improve while time and cost would be reduced simultaneously.

The technology is available today to make drastic changes in the fundamental design process of engineered products. Predictive engineering is a technology that enables companies to change the way they develop, design, and manufacture products. With most of industry moving to time-based competition, predictive engineering will emerge as a major tool in the struggle to regain competitiveness in the design and manufacture of engineered products.

TOMORROW AND BEYOND

The rate of change in the technology revolution continues to increase and shows no sign of abating. The next few years will witness more changes than the last century—a tremendous acceleration. The key to a successful organization will be integrating the new technology into a continuously changing enterprise structure that reflects a commitment to continuous improvement.

Many companies fall into the common trap of creating a structure that makes technology a separate element. This is usually accomplished by creating a senior position such as a VP of Technology. The trouble with this philosophy is that technology becomes a primary means of implementing change. It emphasizes technology for technology's sake, without focusing on the objectives of the enterprise: to produce a quality product in a short period of time at a reasonable cost.

When organizations realize that technology must be integrated into their basic business plans, with business strategy as the primary focus, dramatic continuous improvement occurs. Evolutionary rather than revolutionary, this improvement is a "return to basics." Technology becomes its key enabler rather than a threat.

THE NEXT PARADIGM SHIFT—WHERE TO LOOK FOR EMERGING CHANGES

Changes occurring in computer technology during the past two decades were centered around conventional number crunching and the ability to move data across networks. Today, change is affecting the desktop computer (PC or workstation), impacted by the availability of local supercomputers with their mammoth number-crunching capability at commodity prices, and advances in network technology. Large amounts of data will be moved transparently among locations that may be local or distant, even in different continents. When the three ingredients, workstations, supercomputing, and networking, are integrated with the customer, internal or external, improvements of an order of magnitude can be made to product quality, development time, and the cost of making products.

It is well recognized that network changes will revolutionize the ability to share information. The World Wide Web is here to stay, and it will only get faster and more reliable. The Web, and for most companies their intranets, will support their design simulations with visual catalogs of bookshelf subsystems and components of proven capability. Everyone in a process will then use information technology as an integral part of their work. Just as most people drive without caring how the car's suspension or engine works, the user will not care whether it's with a PC, a workstation, NT, or another system—he or she just wants to get the job done. With already-designed and tested components, the emphasis will be on designing the total product as a system, rather than the pieces that make it up.

Another amazing change that is not in the spotlight is the cost of memory for computer systems, especially desktop systems or departmental servers. Changes in memory cost and size will enable core knowledge and design experience to be stored at a cost of less than $1000. By the year 2010, the cost of storage will be pennies a year—$100 per terabyte is considered a reasonable estimate, and $50 will buy enough space to hold 100 years of design experience.

This means that, just as faster computers allow engineers to perform accurate design simulations quickly, the advances in memory capacity will allow companies to store knowledge of the product and why it was designed in a particular way, and retrieve it later to give others a more complete understanding of the design process.

The availability of knowledge-based information about the product will explode and its successful use and packaging into robust wisdom-based systems will require creativity well into the next century. Companies that continually reinvent their product development core capabilities will not only survive—they will flourish in the next global shakeout.

Visual engineering will continue to expand in functionality. Virtual reality puts the computer-generated concept and its functioning scene inside a person. Goggles, gloves, and other devices are used as interfaces by the user. This technology will become cost effective around the 2000-2005 time frame.

The next step is called *holodeck* technology. The holodeck, in a sense, can be thought of as the inverse of virtual reality. Instead of just looking at a hologram, an engineer could enter a room in which the lasers and computers generate a 3-D holographic setting on the walls. This makes the user feel almost like part of the display. This technology can be used throughout all levels of an enterprise, allowing not only the enterprise, but its customers to understand the design concept. With the holodeck of 2025, *Star Trek* will have arrived!

Index

3-D geometry, 132

A

ACIS/STEP, 96

B

business drivers
    competition, 19
    technology, 20

C

CAD/CAM, 90
CAE, 25, 26, 28, 33, 34
change
    benefits, 17
    dynamics of, 2
    management of, 16
competitive advantage, 29
competitive assessment, 4
computer assisted engineering, (see CAE)
computer-based prototype, 78
concurrent engineering, 86, 118
continued engineering, 71
customers
    external, 8
    internal, 8

D

design communication, 126

E

engineering change
    requests, 31
engineering workstations, 14, 25, 37

F

free-form fabrication, 133

GH

high-performance
    computing, 148
holodeck, 164

I

information technology, 6
installed base, 20
integrated approach, 75
integration, 111
interdisciplinary cooperation, 122

JK

keiretsu, 42

L

laminated objected manufacturing, 136

M

massively parallel process systems, 78, 97

N

neural networks, 101

O

order fulfillment, 140
outsourcing, 103

P

parallel processing, 41
people, 22
people issues, 121
pixels, 98
predictive engineering, 93, 154
process, 22
product cycle plan, 18
product validation, 75

Q

quality function deployment, 66

R

rapid prototyping, 80
reduced cycle time, 31, 32
reduced instruction set computing, 40
re-engineered process, 145
root causes, 10

S

scalable parallel processors, 101, 106
semiconductor investment drivers, 51
software, 53
solids modeling, 72, 95
strategic advantage, 29
systems engineering, 11

TUV

virtual age, 100
virtual desktop, 106
visual engineering, 6, 59
voice of the customer, 64